Pastoral Song

A *Farmer's Journey*

JAMES REBANKS

MARINER BOOKS

New York   Boston

For Helen, with all my love

Originally published as *English Pastoral* in the United Kingdom in 2020
by Allen Lane, an imprint of Penguin Random House UK.

A hardcover edition of this book was published in 2021
by Custom House, an imprint of William Morrow.

FIRST MARINER BOOKS PAPERBACK EDITION PUBLISHED 2022.

Library of Congress Cataloging-in-Publication Data has been applied for.

ISBN 978-0-06-307324-1

22 23 24 25 26 LSC 10 9 8 7 6 5 4 3 2 1

six hundred years. But it is more than a paean to fells (hills), becks (streams), and flocks. Inspired by Rachel Carson's *Silent Spring*, Rebanks's new book urgently conveys how the drive for cheap, mass-produced food has impoverished both small farmers and the soil, threatening humanity's future. . . . What Rebanks brings to this impassioned manifesto for change is a stark time-lapse portrait of his family's land over the course of his lifetime—and eloquent, inspirational ideas for bringing farming back to a brighter future. *Pastoral Song* is a full-throated ode to 'finding a balance' by using the land and animals responsibly and sustainably."

—NPR

"Rebanks, who runs a family-owned farm in England's Lake District and wrote the 2015 bestseller *The Shepherd's Life*, waxes lyrically about his bucolic surroundings while also delivering an eloquent treatise on the problems of modern agriculture."

—*Washington Post*

"Rebanks's lifetime spent farming gives this book its credibility; his sensible tone gives it its power. And his eloquence describing his beloved farm gives it its beauty."

—*Minneapolis Star Tribune*

"Remarkable. . . . A brilliant, beautiful book. . . . Eloquent, persuasive, and electric with the urgency that comes out of love."

—*Sunday Times* (London)

"A paean to a more life-enhancing approach to farming. . . . Rapturous. . . . For Rebanks, farming and writing have proved complementary: while working long hours on the land, he has produced a book in a pastoral tradition that runs from Virgil to Wendell Berry."

—Blake Morrison, *Guardian* (London)

"James Rebanks's fierce, personal description of what has gone wrong with the way we farm and eat, and how we can put it right, gets my vote as the most important book of the year. . . . Written with the raw power of a three-act Ibsen play, Rebanks shows how good people end up doing bad things to their world without meaning to. . . . What makes this book sing is not its argument—which is not unique—but Rebanks's right to say it. He is a farmer. Unlike many dreamier books about nature aimed at townies, he uses hard facts, he manages his own land, and he has fired his potent prose in the furnace of his own experience. But it is more than a polemic. It is also a book full of love: of his grandfather, of his children, and of the Lake District valley where he lives and farms. . . . Some books change our world. I hope this turns out to be one of them."

—Julian Glover, *Evening Standard* (London)

"*Pastoral Song* is a wonder of a book, fierce, tender, and beautiful. Deeply personal but also global in significance, its pages course with love and concern so palpable I more than once wept while reading it. James Rebanks writes lyrically and passionately of the shadow that has fallen over our relationship with land, and how we might reconfigure the ways we think about it, relate to it, interact with it, and with each other. It's both a sobering, urgent read and a deeply inspiring, hopeful one. The book, and author, are to be treasured."

—Helen Macdonald, author of *H is for Hawk*

"Rebanks offers a sensible way to think about food and the planet. . . . His prose will transport readers, introducing them to both the harsh realities and the joys of everyday life on a piece of land that has deep, personal meaning."

—*Christian Science Monitor* (Best of the Month)

Pastoral

Late Middle English, from Latin *pastoralis*

Adjective

1) Of or pertaining to shepherds; hence, relating to rural life and scenes.
2) Relating to the care of souls.

Noun

1) A poem describing the life and manners of shepherds; a poem in which the speakers assume the character of shepherds; an idyl.

Contents

The Plow and the Gulls

The black-headed gulls follow in our wake as if we are a little fishing boat out at sea. The sky is full of winged silhouettes and screaming beaks, and streaks of white seagull shit splatter like milk down onto the soil. I am riding in the tractor, crammed in behind my grandfather. My backside aches from sitting on adjustable spanners, a wrench, a socket set. We are plowing a twelve-acre field, high on a limestone plateau that tilts slightly down to the Eden Valley in the distance. The land is divided into long rectangular fields by silver drystone walls. It feels like we are on the top of the earth, with only the clouds above us. The birds rise and fall in hungry tumbling waves. The highest soar far above the field like children's kites, anchored by lengths of invisible string. Some hang in midair a few feet behind the tractor, wings beating, just above the plow; others glide, motionless, almost near enough for me to touch, with searching eyes and wrinkled yellow legs. One gull floats, with a leg hanging, bent and crippled. The blue-gray Lakeland fells in the distance rise like the silhouetted backbones of giant sleeping dragons.

The three plowshares slice the earth into ribbons, and the shining steel moldboards lift and turn and roll them upside down. The dark loamy inside of the earth is exposed to the sky, the grass turned down to the underworld. The upside shines moist from the cut. The furrows layer across the field like sets of cresting waves sweeping across some giant brown ocean. The freshest lengths are darker, the older ones fading, lighter colored, drying and crumbling, across the field. More seagulls arrive, hearing a rumor blown on the winds to the four corners of the sky. They come across the fields and the woods on eager wings, on flight lines so straight they could have been drawn on a map with a ruler. They scream and cry out to one another, excitedly, spotting the freshly turned soil.

The tractor engine works hard, oil-black smoke spewing from the exhaust, as we head up the hill. My nose fills with the smell of diesel and earth. My grandfather turns backward and forward, half-focused on the straightness of the furrows, using two landmarks ahead, far beyond the headlands, to guide his line and keep it honest. One mark is an old Scots pine, the other a gap in a wall on a distant hill. He tells me about a young plowman he knew who used a white speck as his more distant sightline mark, but ended up with crooked work, because the farthest mark turned out to be a white cow that was walking to and fro across a distant hillside. The other half of my grandfather's focus is on looking back to ensure the plow does its work behind him. So he sits half-twisted between the two angles, the muscles in his neck taut, his leathery cheeks rough with silver stubble.

The gulls fall upon the virgin soil and grab worms from atop the loosened surface. And then they quickly take to the sky again, racing away, in a mad wing-flapping dash, gulping down their catch as fast as they can before they are mobbed. When they have the feast stuffed safe in their bellies, they are a hundred yards or more behind the plow. They flap back into the air and gain height, and glide down the field until they are above the tractor again, and then they repeat the whole cycle, over and over. Farther down, the rooks march across the field, and some of them take to their black wings and join the swirling crowd.

There is a groan as metal scratches across the limestone bedrock. The tractor suddenly strains, engine toiling, like someone has dropped an anchor, then metal creaking, and stone breaking, and the plow lifts a little and surges forward, released. A slab of rock appears behind the plow, sprung to the top. The biggest stones remain largely submerged, like icebergs, just the scratched tip, or a broken-off fragment, showing above the furrows. The soil on this hard farm is shallow, so this happens again and again.

The night creeps in. The shadows lengthen. The seagulls head off for their roosts in giant Vs. They look to me like the bomber formations in war films. The fells tremble and flicker in the darkening blue light. The headlands are plowed, the work is done. And we head home. The tractor headlights shine a halogen-yellow tunnel through the branches that arch over the road. Rabbits scurry across in front of the tractor into the verges. I sit, yawning. Fat white stars flicker in the blue-black sky. As the tractor travels back through the little village, the houses are glowing with electric light, TVs and people walking about in their kitchens or slumped in their living rooms.

~

Every journey must start somewhere, and this is where mine began. I sat in the back of that tractor, with the old man in front of me, and for the first time in my life thought about who we were and what the field was, and the relationship between the gulls and the plow. I was a boy living through the last days of an ancient farming world. I didn't know what was coming, or why, and some of it would take years to reach our fields, but I sensed that day might be worth remembering.

This book tells a story of that old world and what it became. It is the story of a global revolution as it played out in the fields of my family's two small farms: my father's rented farm in the Eden Valley, which we left nearly two decades ago now, and my grandfather's little Lake District fell farm, seventeen miles to the west, where I live and work today. It is the story, warts and all, of what farming was like here in my childhood, and what it became. It is about farmers like us, in our tens of thousands, across the country and around the world, and why we did the things we did—and what some of us are now trying to do to make it right. The last forty years on the land were revolutionary and disrupted all that had gone before for thousands of years—a radical and ill-thought-through experiment that was conducted in our fields.

I lived through those years. I was a witness.

NOSTALGIA

The hardest thing of all to see is what is really there.

J. A. Baker, *The Peregrine* (1967)

A healthy *farm* culture can be based only upon familiarity and can grow only among a people soundly established upon the land; it nourishes and safeguards a human intelligence of the earth that no amount of technology can satisfactorily replace.

Wendell Berry, "The Agricultural Crisis as a Crisis of Culture," in *The Unsettling of America* (1977)

We sit silently in the waiting room, perched awkwardly, like nervous crows, on the stiff-backed chairs. Formal portraits of the founding fathers of this law firm look down sternly from the walls. Seated beside us there are a slightly graying mother and her daughter. The daughter whispers to the mother and she whispers back. Then they are ushered up the stairs by a man in a pinstriped suit. These stuffy Dickensian offices are beside the sandstone church in our local town. The steps up to the door have been worn away by the best shoes of generations of country folk scurrying in and out to sort out various legal issues.

The first mention of my family on paper concerns a legal dispute about land ownership with a local aristocrat in 1420, in the neighboring parish. We are here at the solicitors that has handled our farm's legal affairs for at least three generations, to learn the details of my father's will.

My grandfather's solicitor was simply spoken of as "Charles," as in "We'd better ask Charles about that," when anything remotely legal came up. Little market towns like ours have long had a smattering of middle-class professionals who serve the needs of farmers and others who live from the land.

A young woman who seems to be a trainee secretary offers me a cup of coffee. An older woman has prompted her, with a nudge and a whisper, to ask me, but it soon becomes clear the young woman doesn't really know how to work the coffee machine. It seems she is trying to do her best in a new job but has yet to find her feet. Her hands tremble with the cups. "I'm not a posh coffee person," she says to us under her breath, embarrassed. The older woman moves her gently but firmly to one side and makes the coffee. The young woman, now back behind the desk, looks as if she'd like to run away. I know that look. Until I was in my twenties, I was terrified even of having to chat with "posh" people (anyone vaguely middle-class or university-educated). I felt small around them, or else I simply turned surly and quiet. They had all the words. They knew all the things I didn't.

No sooner has the coffee arrived than we are politely escorted by the older woman across the corridor and into a room with leather-upholstered chairs around a varnished table. Through the window, beyond the table, two gray pigeons strut after each other on top of a slate roof. A woman enters the room behind us and passes my mother with an armful of old and bulging folders tied with string and ribbons. She makes her way around the table and introduces herself, and tells us that these are the "deeds" for our land. The ribbons are untied and the bundles slouch and spread like a fat man's belly released from a belt. I long to open out these papers, this thick wedge of untold stories, and hold them in my hands, but clearly not many people ever do that here because she tells us the legal necessities we have come to hear and the deeds remain spread loosely, but unopened, on the table. The solicitor starts to speak, but I don't hear her words. She sees that I am distracted and pauses. I ask if I can look at the deeds. She says I can. She pushes some of them toward me and begins to explain. The first two or three documents are open in our hands at their stiff folds, like giant cardboard butterflies unfolding their wings.

In these pages is the nearest thing to a written history of our land that exists. The waxy sheets are spider-scrawled with almost illegible copperplate handwriting and pastel-shaded sketches of fields. Giant antique letters open each crammed page. Melted burgundy-red wax stamps are surrounded by earnest signatures. As my eyes become accustomed to the script and field sketches, I see a half-familiar world opening up of field names and landscape features—trees, becks (streams or smaller rivers), lanes, and barns—a parallel paper-and-ink negative of the grass, stone, soil, wood, and landscape that I know. There are historic features I have never seen before, like archaeological finds, all marked "Celtic."

The history of the ownership of every field is in these bundles, and every transaction is detailed in them, going back centuries. The last time these were seen must have been by my father or my grandfather, and before that by the people who farmed here before we did, because the deeds have been stored away from our grubby hands, in the archives. They were only consulted when there was a dispute about a boundary or ownership of some place or item or other, or when somebody died. The field names catch my eye:

Greenmire
Little Greenmire
Smithy Brow
High Stoney Beck
Clovenstone
Cloven Stone Rigg
Browfield
Wood Garth
Long Field

Somewhere in this bundle of deeds is the transaction for my grandfather's purchase of a hundred acres here in the early 1960s. He had taken my father, then a skinny teenager, and his brother-in-law Jack, who knew that country better than he did, for a Sunday afternoon ride out to "see something." He drove them to see this little run-down, scruffy, badly fenced, scattered collection of fields, mentioned in these deeds, that together made up a "fell farm." He declared he was going to borrow the money and buy it for summer grazing for his cattle and sheep. It cost £14,000. There is also the paperwork for my father and mother's purchase of fifty acres in the middle of those fields from another retiring farmer—to make it a whole farm—and later the addition of a farther sixteen acres in the 1990s when the adjoining land came up for sale. Soon this archive will contain the deeds for the fourteen acres my wife and I bought up the lane behind our house in the weeks after my father's death, because they are near to our farm and will be useful for our sheep and cattle.

These deeds show land passing from one family to the next, again and again, and remind me that a farm isn't a fixed thing but often changes with every generation, as families buy or rent, or sell land. This history is messy and complicated, like that of most families. People's attachment to their land is renewed by each generation—through their holding on and working it. It could also be lost. As the solicitor speaks, I know that my family's future on this farming landscape in a corner of northern England will be determined by my ability to earn enough from our land (and any other way I can think of) to pay our bills, service our debts, and make some money for us to live on. Ever since I was a teenager I have worked on our farm, and been the shepherd of a flock of sheep, but this is different. When we walk back down those worn sandstone steps of the solicitors, I know that I am now the "farmer."

~

The months after my father's death were the hardest of my life. I had always wanted to be the farmer, the captain of the ship with my hand on the wheel, but the moment it happened it felt empty. The world seemed a dull shade of gray. Beyond our little valley, people everywhere seemed to have gone insane, electing fools and doing strange things in their anger. England was divided and broken. Suddenly in those months I felt lost. It was as if I had been following in someone else's footsteps down a path, talking to them, reassured by them when the going got tough, and then they had disappeared. The farm was a lonely place—a poorer thing when it wasn't shared. And with every passing year farmers were becoming fewer and fewer, a vanishingly small and increasingly powerless share of the population. Our world felt fragile, like it might now break into tiny pieces.

~

The UN says that 5 million people move from rural communities to urban ones every month, the greatest migration in human history. Much of this took place two or three generations ago in Britain, the "first industrial nation." So ours is now one of the least rural societies on earth. The majority of people now live in towns and cities, and we tend to give little serious thought to the practical realities of farming, the vital moment when we come up against the natural world.

And yet we are all still tethered to the land in a practical sense—our entire civilization relies on farming surpluses, which free most of us from growing our own food, allowing us to do other things. We are no longer the slaves of the "dark Satanic mills" of the industrial era, but millions of us are still reluctantly chained to desks in the soulless corporate offices that followed. We act as if we popped into town to earn a living a generation or two ago, but will be going home soon to a place in the country. There are few things we profess to care about more than our most beloved landscapes, or "nature"; few dreams more enduring than finding our way back to the pastoral of small villages, farms, and thatched cottages, with little fields edged by hedgerows smelling of honeysuckle.

They used to call England a "green and pleasant land," but in truth it was never entirely green, nor entirely pleasant. It was a tough old place with almost every acre used by humans, but there was much in it that was good. And yet the truth is that the countryside that feeds us has changed. It is profoundly different from even a generation ago. The old working landscapes and the wildlife that lived in them have mostly disappeared, replaced by an industrial farming system that in its scale, speed, and power is quite unlike anything that preceded it. This new farming has proved to be both productively brilliant and, we now know, ecologically disastrous. The more we learn about this change, the more unease and anger we feel about what farming has become. Our society was created by this farming, and yet we increasingly distrust it.

This was a lousy time to inherit a farm. I was now solely responsible for making the decisions about how we managed my family's land. In the months after my father's death five years ago I began to feel a kind of despair. Our role was now being challenged and criticized as never before. Reports of bad news and scientific studies about the decline and loss of wild things on farmland became commonplace on the TV and radio. Rain forests were burned, rivers poisoned, soils eroded, and countless landscapes made sterile and bereft of nature. Anger filled the newspapers and the news. Being a farmer felt for the first time like something you were supposed to say sorry for. And with some sadness and shame I could see that there was truth in all this. My new role wasn't heroic, as I had imagined it would be in my youth. It was just confusing and complicated, and fraught with doubts. Countless choices—some large and fundamental, and others tiny, incremental, and day-to-day—that would shape this little bit of England for better or for worse were now mine to make. It felt like a lot rested on my knowledge, or lack of it, and my values and beliefs. And I was suddenly aware of how constrained my choices were, and how little I knew. I would have to work out how to make money from our land without wrecking it. I had inherited a complex bundle of economic and ecological challenges— and that, perhaps, was what it really meant to be a farmer.

When we lose our way, it often pays to retrace the footsteps on our journey until we get back to familiar territory. In those painful first months, my grandfather's farming became for me such a moment from which I could navigate through what had happened in order to understand what had gone wrong. I thought a lot about how he managed his land and cared about his animals and the natural world around him. I tried to understand afresh what it meant to be a farmer. I returned in memory to a day spent plowing a field in April, nearly forty years ago. Every detail was frozen in my head. Forty years doesn't sound long ago, but in farming terms it is like returning to the age of the dinosaurs. Perhaps I would only discover old mistakes, or get a nostalgic sense of what that traditional farming was. But I returned to the past with a sense of hope, that it might hold some of the answers—and help me to work out what kind of farmer I could, and must, become.

~

As I sat in the back of his tractor watching the seagulls, it felt as if Grandad and the seagulls behind his plow were part of the same whole, the one as true as the other. They both had timeless claims on the earth; they both belonged to the same cycle in that landscape. They needed each other. I was aware, perhaps for the first time, with absolute clarity that we were farmers and that defined us beyond anything else. We changed the earth to grow food so that we, and others, could live. My grandfather was rooted in work, connected to the soil and the crops and the animals upon it. I loved his closeness to the land. I was dimly aware that lots of people didn't live like us. Most families, even in our village, had traded in their relationship with the land for new lives away from the fields, birds, and stars.

Earlier that spring my grandfather had decided it was time for my farm "education" to begin. He set out to teach me the ways of his world. I had perhaps always been vaguely aware of the cycles of the work, as I had trailed after the men since I could walk, but this was different. He had sensed in the preceding months that the farm was losing me. I was work-shy and beginning to hide in the house, huddled by the TV. He knew that I would either learn to love farming now or drift away and be lost from it forever. I was old enough to be pried away from the house and the women—to start learning and be useful. I didn't get on well with my father, and that was poisoning my view of the farm. I would try to help him and would inevitably do something wrong and be shouted at. He seemed rough and best avoided. Skulking inside was easier. But I felt ashamed, because I knew this wasn't the boy I was supposed to be. I was in danger of becoming a disappointment.

~

The old farmhouse glass had imperfections in each pane— whorls, like knots in an oak-tree trunk—which distorted the sycamore tree in our garden, the clouds, and the electricity pylons. I thought the fields were featureless and bland. I was a daydreamer. Then my father bawled at me to get my boots on and go out and help, shouting that he wasn't running a holiday camp. When I appeared at the back door, he told me what jobs to do outside and turned in disgust back to the yard, leaving a shitty brown stain dripping down the cupboard, where he had been standing. And all I could think was, who would want to be out there, with frozen hands, working for that maniac in the rain?

~

One day I heard my grandfather shouting at my father that he had made me "sick" of the farm already, that he had been too hard with me. Grandad was still the patriarch of every one of our acres, rarely away from the fields, and I soon learned that it was more fun working with him or, even better, being on his farm up in the fells, than working for my dad at home.

Grandad didn't look much. He wore the same old brown suit every day. Beneath his flat cap his scalp was pale, and his hair swept across it pathetically. He had a cup of toothpicks by his chair and would use them to fiddle for bits in his teeth. He had never seemed young, looking more or less the same, but skinnier, in the oldest picture we had of him, in which he held a prize beef Shorthorn bull in front of the castle in the local town. But I didn't care what he looked like. I seized any chance to hang out with this old man who told amazing stories and seemed to do whatever he liked. That year he set out to teach me all about the fields, starting with the plowing of the barley field and how our farm worked through the seasons. He knew that given a little time with me he could make me fall in love with it all. And he was right, because in the course of that year I did. Nearly forty years later, that year mattered in another way, because it gave me a head full of memories of a farming world that would soon vanish. That time became a lifeline for me, a light in the darkness.

My farm education that year was fragmented, handed to me like pieces of a jigsaw that didn't always form a whole. These fragments would only slowly build up to give me a clear understanding of that world and its values. I was learning the old ways, and just in time, because they were starting to die out all around us, even in our own family. I had uncles and cousins with good lowland farms fifteen miles away, and it was clear from their new tractors, machinery, and big buildings, and their barely concealed contempt for our old-fashioned farming, that things had already changed for them.

~

He sat in his Land Rover in the farmyard, by our back door, revving the engine and beeping his horn. My mum said I'd better get a move on or Grandad would go without me. I tumbled over my own feet, trying to put my Wellington boots on and get through the door. I was to be Grandad's "gate-opener." We rattled off down the lane, with him grumbling about being late. A minute later, he stopped at the Long Meadow gate, and I jumped out and opened it quick-sharp (only the heaviest gates, or those tied with barbed wire, got him out of the vehicle). He drove through and I closed the gate behind him.

Some of the pastures were full of ewes and young lambs. He took care to see that they were all being mothered correctly and were thriving. He knew which lambs belonged to each ewe by sight and could tell when one was missing or following the wrong mother. We drove around his "stirks" (young beef cattle) that had just come out of their winter barns to graze. These young cattle were flighty, raising their heads and galloping off, snorting like startled wildebeest. Grandad said they were fine, and we needn't bother them. Farther down the road three escaped lambs galloped along, baaing for their mothers, and tried to push themselves back through the hedge. Grandad traveled with a bucket of staples, a hammer, and a roll of wire in the back, for moments like this. He sent his sheepdog Ben to get the lambs back while he patched up the fences. We let the flock through a gate to a new field. He said they were "a bit stale" and that my father should have moved them already. Sheep should not hear the church bells twice in the same field, he said—it meant they had been in one field too long.

He parked the Land Rover, and we walked across a sandy bank covered in gorse to inspect more of our fields. I tried to keep up with his strides, just as I tried to pee for as long as he did when he emptied his bladder (but I couldn't because he pissed like an old horse and it went on forever). As he walked, the grass rasped at his boots, making a scythe-like sound with each stride. He wore old brown leather boots that turned up at the end like clogs, with yellow laces, and my grandmother polished them with dubbin. Halfway to the cows, he paused to take in the wider panorama of the valley, reading the varied greens and browns, the patchwork of different grazing, and the other farms. He knew exactly what everyone else was up to in the valley. As we walked, I was told that every one of the crops and animals had its own annual cycle of birth or planting, growth, protecting and feeding, harvesting or killing and selling. It would be another ten or fifteen years before anyone told me the technical term: an old-fashioned "mixed" and "rotational" farm. It didn't warrant a name to my grandfather, because it was just how everyone he knew did things.

~

As we walked the farm, a bewildering—to me—number of different things were happening in the fields around us. There were four or five meadows growing hay, with a few for silage, and two or three of barley (including the one I had just helped him plow). Down the headlands, there was a field of freshly sown oats, grown for the horses; then a field of turnips for the sheep; and a dozen or so raised mounds of soil, "stitches," planted with potatoes for the house. Farther away, a field of "whole crop" of peas and beans, yet another source of winter cattle feed. As if this wasn't enough, he also—reluctantly, but to keep my grandmother happy—had a garden with rows of cabbages, lettuces, carrots, and onions. He would curse and grumble as he broke the earth into clods with his fork each spring. He had, until recently, even more kinds of livestock in the fields and barns: a herd of dairy cattle, another for beef, three breeds of sheep, and pigs, horses, and hens for eggs, and ducks and turkeys reared and fattened for sale at Christmas. To my eyes, my grandfather knew how to grow and care for a lot of different things—he was a farming jack-of-all-trades.

He told me one day not to be confused by it all, that the pattern was simple. "The farm dances around the plow," he said. Other tools followed and were used, but the plow was king. To grow a crop, he had to plow, or "till," the ground to create a seedbed, turning the shorn old crop into the earth to stop it from growing. The plow was the key tool for "improving" his rented farm—and had been since he'd been a young man in the 1930s and 1940s, plowing with a horse and trudging up and down the furrows in his hob-nailed boots.

There was something about working the land on foot behind a horse that seemed to make him see the world differently from the way later generations would see it from a powerful tractor. My grandfather knew our fields as if they were extensions of his body. He had felt the plow tremor as it scratched across the bedrock, felt it in his hands and through his boots. On foot, behind a horse, grass, soil, and worms were up close—seen, heard, smelled, and touched. There was nothing between him and the nature he worked with. The labor was often hard, long, and perhaps sometimes boring, but I never heard him say a word to suggest he regretted a minute of it.

I walked and rode with him through the seasons—watching and listening. In the midst of Thatcher's Britain, I was a boy on a tractor listening to my grandfather's tales from the 1930s (or others from the 1890s that his grandfather had told him). These tales were full of horses. There was a kind of magic in those stories, because the horses and men in them were already gone. The sun was beginning to set on his world. He was in his final field days.

~

We followed the old drystone wall that led from the farmstead to the meadows. The wall rose and fell with the contours of the rigg, the raised hogback mounds of earth that rose above the floodplain. Meadow pipits rose up from the field and flitted up and away ahead of us, perching on the posts that held the wire to the top of the walls to stop the sheep from escaping. The ewes and lambs called to one another across the valley. My grandfather stopped and held his hand to his ear, pantomime-style, to listen to the cuckoo calling in the woods on the fell. I nodded. He quietly opened the wooden gate by the stone barn, or "hoggust," as he called it. An old black Aberdeen Angus cow was due to calve. He had walked her into the barn the previous night, so that he could help her if things went wrong.

We peered through the broken window. A jet-black calf was lying in a patch of light flooding in over the barn door. My grandfather crept in quietly and I followed, pausing at the door. The cow's teats were shining and soft, so he knew the calf had suckled. It shone, in wet cow-licked curls of saliva and spittle. The old man talked to its mother to reassure her, and she lowed back at him, but then relaxed and let him scratch her rump. He pulled gently at the afterbirth, and it peeled away and fell in a sloppy bundle on the barn floor. He lifted it away with an old pitchfork and threw it into a clump of nettles by the blue slate and lime mortar wall of the barn. He reached under the calf and said it was a bull calf; then he lifted it to its feet. The mother watched him carefully with big watery black eyes, chewing her cud. He waved to me, and I knew he meant that I should open the door. The cow wandered through, calf stuttering after her, and away across the meadow to the rest of the herd. She paused every few strides to let her wobbly-legged son catch up. We watched them, the cow heading to the beck for a drink, grabbing wisps of grass as she went. Some of the other cows came to inspect the calf and touch noses with the mother. The rest were grazing across the field, tails swishing as they munched along; others were lying with their calves by their sides and flicking their tails and ears at the flies that encrusted their flanks and eyes with shimmering specks of emerald and black. An older calf was gently pummeling his mother's udder as he sucked, face covered in a milky froth, while another stole milk by reaching in from behind while she daydreamed. Grandad turned to me and said, "That's a cheeky bugger, isn't it? It thieves milk off all these old girls when they're not looking; no wonder it is fat as butter."

~

Time seemed to slow down around my grandad. He believed in watching carefully and taking time with his animals. He would simply gaze at his cows or sheep for what felt like ages, leaning over a gate. As a result he knew them all as individuals. He could spot when they behaved differently because something was wrong, when they were coming into season or were about to give birth. He thought only fools rushed around. He believed that a good farmer was patient and used his, or her, eyes and ears, and nose and touch. Doing things well was his goal, not doing things quickly or with the least effort. He called me his "squire," which I never understood until much later—I was his project, his apprentice. A week after we plowed the barley field we went back to "pick stones." My grandfather didn't spell it out, but I see now that that field was to be my classroom, the place where I was to learn all the stages of growing a crop.

~

The furrows had dried and crumbled in the wind and sun. I was told to steer down the eight-acre field in the lowest gear. I bumped along on the tractor, crawling slowly down the furrows, getting bouncier at the front as the weight in the back grew, stone by stone. My grandfather and John, a bandy-legged farm worker with Brylcreemed black hair and blue cotton trousers, walked behind, throwing stones into the "transport box," a metal crate suspended on the back of the tractor on the hydraulic arms. Each fist-sized rock the men threw arced through the sky and clanged into the box, or cracked as it settled on the other stones. As I began to worry about crashing into the wall at the field's end, Grandad climbed back on, nudged me out of the way, and took the wheel. He drove the box of stones away to tip into holes in the field, or in gateways or lanes to make them firm. Any good walling stone was taken to where it could be reused. Nothing was wasted. Stone was a useful thing. And boys were meant to be useful too. The truth is, I was a lonely kid, awkward and easily embarrassed. Other people made me feel nervous and as a result I'd do or say stupid or clumsy things. But my grandfather was different. He made me feel respected and important. I would do anything to make him proud of me, so, when he began my field education, I paid attention, even though I wasn't really sure I wanted to be a farmer.

~

After the stone-picking, the next day we had to break up the furrows into a seedbed. The harrows were big upside-down iron rakes, the size of a couple of double beds, held together by chains and pulled by the tractor over the furrows. With each clattering pass, they slowly broke the ground into a crumbly flat surface. After a while, Grandad said the seedbed was ready because you could see the harrow lines left in the finely raked soil, like fingers pulled through dry sand. My father appeared at the top of the field with the seed drill, an ancient-looking contraption that sowed the grain in the ground at periodic intervals, somehow dropping a single grain every three or four inches or so (or so we hoped, because it was a waste of time if it didn't). He drove past and mimed "You OK?" I nodded back.

Three generations of us. The field a blur of tractors and dust. And with each pass the work got done.

~

A week later, the soil glowed warm in the first days of sunshine. Then we were rolling the field, flattening the loose soil, tucking the seed beneath a pressed-down surface, away from the rooks that sought to steal it. Or, rather, my grandfather was rolling the field; I was being anywhere but at home with my dad, who was in a foul mood because a calf had died of scour. Behind us the giant iron roller rumbled along, the huge cylinder full of water clanked, rattled, and banged every time it hit a bump in the field.

I bounced around, daydreaming about the John Wayne movie I had watched that morning, the one in which he hires a bunch of schoolboys to do a cattle trail (because the men have all disappeared to the gold rush). He is then killed by some thieves, but it is OK because the boys get tough and avenge his death by hunting down and murdering the bandits.

My grandfather was saying something or other about the peewits (also known as lapwings) that flapped around us with their paddle-like wings, twisting and turning, rising and falling, flashing their wings. Suddenly, he stopped the tractor, climbed off slowly, cursing his stiff old legs as they planted themselves in the freshly tilled soil. He strode across the ground, eyes fixed on the spot. I wondered what he had seen. He bent down and picked something up from a scratch in the ground and put it in his flat cap. Then he climbed back in and set the cap on my knee. I looked down at the eggs and held one in my hand. It was warm, and the mottled color of the boiled imitation pebble sweets you could buy at the seaside. They are curlew eggs, he told me. They nest in these fields. We bounced on. When we had come full circle, he took the cap full of eggs, climbed out, and placed them back on the ground where they were, re-creating something like a nest with the back of his knuckles. I asked him if the parent birds would come back to them, and he said, "Sometimes they do, sometimes they don't . . . but this is the best we can do."

When we came around the field ten minutes later, the

mother curlew was nestled in the dusty seedbed like nothing had happened, and my grandfather grinned. That night I told Dad proudly about Grandad and the curlew eggs. He said Grandad was a "soppy old bugger" and no wonder we had taken so long to get the work done.

~

Two weeks passed and the barley was poking up through the earth, the little spears of green racing for the sky. There was a notable sense of relief as my grandfather walked across the field, through hundreds of regulation rows of green seedlings. While I was at school, my father spread some artificial fertilizer on the field; I could see it like tiny little white polystyrene balls on the ground. And with every passing day the curlews, oystercatchers, and peewits on their nests slowly became submerged in a rising sea of nitrogen green barley.

~

My grandfather rarely went to church, and thought the vicar was an idiot, but he would say things like "you'd better pray" when I asked him whether the barley we had sown would grow. Planting a crop was an act of faith. There was a very real sense that it might all end in failure. The seed drill might not have worked. The birds might steal the seed. The weather might be too wet or too cold, or drought might ruin the crop as it grew. And even if the seed germinated, it could be blighted by disease or eaten by vermin, making all that work futile. Everyday disasters of this kind would leave our farm short of crop for our animals in winter.

The year before, our barley had been damp when we harvested, because there wasn't the weather to get it dry. It had warmed in the loft, like a compost heap, until it steamed. In midwinter, when the cows needed it for feed, it had become stuck together and moldy. My father said they would have to eat it or go without. The cows looked at it with disgust. I was beginning to understand why everyone we knew moaned continually about the weather. We were slaves to it.

A field with a healthy, weed-free, and bountiful crop was always what we hoped for, but it was not a natural thing, and if it happened at all, it was created by the will and toil of the farmer. The gods could either reward us with a bountiful crop or crush us in countless ways. And such a hard way of living made for hard people.

~

We walked across the headlands of the barley field to the sandy banks that were pockmarked with rabbit warrens. It was three or four weeks since the barley had been sown, and now the damage done by the rabbits was showing. The first hundred yards from the dike, nearest the warrens, was grazed bare—nibbled off at ground level. Five inches beneath where it should have been. My grandfather had told John to "do something" about the rabbits or, he said, there would be no barley to harvest later in the summer. And I followed him because he was patient and kind, and he taught me things. He lived with his wife, Sheila, in one of the council houses below the barley field. My dad said John was "wasted" on rough farmwork. He was careful and steady, a craftsman at heart, with pride in his work however simple the task. He could make or mend things and took great care in laying bricks or blocks straight and true. He could fashion beautiful gate-catches out of bits of chain, wire, and old nails.

Behind their house, by the coal bins, were two cages with ferrets in them for "rabbiting." The ferrets ate from old Fray Bentos pie tins. Stiff brown rabbits were hung up by the back door, ready to be skinned. John told me not to put my fingers through the wire mesh front, or the ferrets would bite and not let go. He reached in, grabbed them confidently by their middles, and stuffed them into his wooden box and closed the lid. He threw the leather strap over his shoulder and made off through a hole in their fence. I tried to step into the impressions his boots made in the soil, but couldn't quite match them and had to leap a little between strides. A few hundred feet ahead of us a slow creeping tide of rabbits moved, in stuttering waves, back to the stinging nettles around their holes.

At the warren, John did a recce of holes as if trying to

work out a puzzle. He kicked at the nettles to clear them from the holes, then methodically covered their many exits, gently splaying nets made of soft white string across the shadowy holes, like a spider's web. Each net was surrounded by a drawstring attached to a hand-carved wooden peg pushed firmly into the ground. He lifted a ferret from its box and slipped it under the net, feeding it hand over hand into the hole. Then we waited. I knew John was anxious, because the ferret could kill down the hole if it trapped a rabbit in a dead end, and might not come back. He had a spade to dig it out. The ferret's job was really to spread panic among the rabbits, scaring them up into the nets. John watched the nets like a hawk. After maybe twenty seconds, a fleeing rabbit trapped itself in the pursed net. It was caught fast and lay silent, wide-eyed. He swept it up into his hands, peeling it quickly from the net, and pulled its neck and its back legs apart until there was a gristly crunch. It shuddered and went limp; then he tossed it onto the grass by my feet, where it was soon still. He quickly and carefully reset the net. Two other rabbits bolted out of a hidden hole he hadn't netted. He cursed. Then two others were caught in a net and killed. A little while later the ferret emerged, and was caught, hanging limply over his hand, a slightly savage grin on its face. John put it away in his box, and we walked home across the half-ravaged field with the three rabbits swinging from his knuckles.

~

Many years after we caught those rabbits, I read the Roman poet-philosopher Virgil and realized that my people belonged to an ancient farming tradition. Virgil wrote a curious little book, two thousand years ago, called *The Georgics* (which loosely translates as "farming things"). It is a kind of handbook for how to be a good farmer. In it Virgil lists the modest tools (or "arsenal") available to the Roman farmer: plowshare, plow, harrow, horse-drawn cart, threshing sledge, horse-drawn sledge, iron bar, hurdles, and winnowing fan. Virgil said the farmer must use these tools to "wage war" on the earth. His farming philosophy was that we had to take things from nature by using our wisdom and our tools, because the alternative was defeat and starvation.

If you harry not with tireless rake the weeds,
if with your voice you do not terrify the birds
or with your sickle prune the canopy shading the land,
if with no prayers you call down rain,
O! how you'll gaze in vain at another's ample stockpile
and shake the forest oak to soothe your famine.

I didn't know anything about Virgil as a child, but I did sense then that the rabbits were one small battle in an endless war. A war we had to fight, a struggle without end.

~

We heard them down the field. Kraak, kraak, kraak. I knew the noise and what it meant: corpse-chatter. As I followed him down the field, he saw the birds hopping up and down from some rocks onto something unseen beneath a little thorn tree. My grandfather cursed under his breath. He hated losing sheep. Hated what crows did to the dead or dying. He said Mother Nature was a "cruel old bitch." The crows saw him coming and hopped onto the barbed-wire fence, and then over to a nearby oak tree to watch us.

We found the old ewe lying on her side, kicking. She had mastitis, a swollen udder, and it had got into her body. Grandad showed me that the infection had spread up the swollen milk veins into her body. There was no hope. I could see the blood on her face—the thin bright red of squashed strawberries against her white wool. The crows had stabbed out her eyes when she couldn't get up. Her month-old lamb watched from about twenty feet away and then ran off down the field. He said we would have to catch it tomorrow morning in the pens, by bringing the flock in. But the ewe was in pain and blind. He said that if we left her and went home for a gun, the crows would come back and torment her, and she would suffer horribly.

He gestured for me to stand back, then pulled out his knife and sharpened it on a stone. Then he held the ewe's head and cut her throat with two quick slits. I thought I heard him say sorry, but so quietly I couldn't be sure. The blood gushed out, a hot purple-red river pouring down her opened neck into the meadow. She shook a little, legs thrashing, then slowly breathed her last. She was done. Grandad said we would come back for her corpse in the morning. Until we got back, the crows would have their way, but they couldn't hurt her any longer. He shouted at them. Fuck off. One of them lifted off briefly from its branch, as if listening, before settling again. The crows and my grandfather were old enemies, and as the weeks passed, and I saw what they could do to our work, I inherited his fury against them.

~

My childhood world had become filled with countless cycles of birth, life, and death. My days with my grandfather were filled with helping animals to give birth, or stay healthy, or get enough to eat to cope with the weather. At times he could be gentle, with moments of great tenderness and care—nestling a newborn lamb in his hands, gently threading the stomach tube over its little pink tongue and down its throat and squeezing milk into its belly to save its life. But at other times, when he thought it necessary, he could be a tough and almost cruel man. There was a hardness in him that he was neither ashamed of nor uncomfortable about. To him, death and killing were simply part of his life. At the same time he had strong ethics. Even if an animal was going to be slaughtered tomorrow, it was still our job to do everything in our power to keep it alive and well cared for today. Anything other than treating our animals decently and with care was considered wrong and shameful, and a waste of a life as well as time and effort. There was a time to live and a time to die. When he killed, he did it swiftly, with respect, but without great displays of emotion. Knowing and seeing death on personal terms, he had a kind of reverence for meat on the table. We were told not to leave a morsel, even the bacon rinds. He would have been confused that anyone could be so foolish, or rich enough, to suffer rabbits destroying a crop, or so morally elevated to think they were above killing when it was called for. He existed in nature, as an actor on the stage, always struggling to hold his ground. A risen ape, not a fallen angel.

~

In May the hawthorn dikes around the barley fields frothed white with blossom and hummed with bees. The swallows hunted under the overhang when it rained. In the pastures, bullocks scratched their backs under the branches, or stood in their shade, to shelter from the sun. Late spring meant lots of work, but good work, Grandad said, because it achieved something, unlike the dull repetitive daily slog of winter that I'd known and disliked a few short months ago. The ewes and lambs were tagged, marked, tailed, and vaccinated. My father and grandfather had time to breathe a little, and do odd jobs, making good the damage of winter.

It was one of those May days when the cotton-wool clouds raced to the distant fells and the emptying sky was blue. My father had gone to a distant auction mart to see "what sort of trade sheep are." My grandfather told me he had some "gaps to put up," so I went with him. Following him around the edges of our farm, I felt we were like an ancient tribe walking the boundaries of our territory.

We walked across three fields and then down to the bottom of our "Long Narrow Fields." A wall had come down where the winter weather or the sheep had moved a stone or two and weakened it. Some of the stones had tumbled down the bank on the far side. I was sent to carry them back up the bank one at a time. If we left them in the hay meadows, he said they would do great damage to the machinery. The mower would jump and shake with shuddering bangs, as if it had swallowed a land mine. And it had to be done before the grass got too long. My grandfather worked out how best the stones would fit back in the hole like a jigsaw puzzle. He carefully sorted the walling stones, placing the top stones to one side, and divided the good walling stones carefully between the two sides of the wall. Then he began placing them, taking trouble to present their moss- and lichen-covered faces to the outside edge and leaving a flat surface on top of them that could be built upon.

Among the stones at his feet were a little broken clay pipe and an old green bottle, revealing that other men long before us had done this work. He told me stories about his grandfather, and how he became a successful farmer, and about his Model T Ford, and the gold watch he gave his daughter as a reward because she lambed his sheep and was better than any of the hired men. And his landlord's son who, he laughed, once got a "real job" driving a concrete mixer, but had stopped at a pub and got in with a bad crowd, and when he came out, drunk, hours later, the concrete had set solid in the back.

His stories were full of instruction about the kind of people we were and the kind we were not. He impressed upon me that our side of the wall was where we had to work to prosper. He said a farmer was as good or bad as their farm, and what was in it. If our fields swelled with great crops of barley, turnips, or hay, or were grazed with fine cattle or sheep in deep, lush green grass, then we would be judged "good farmers." If they were badly drained and rank with weeds, with tumbling-down walls, badly thriven sheep, or wormy cattle, then we were more or less "losers."

As he laid the top stones on the wall, an hour or two later, I sensed him distracted by something. Instead of lifting his stone, he turned his gaze down the scraggy overgrown headland. As I worked, he touched my hand to still me, and then touched his own ear as a signal to let me know he had heard something. "What?" I whispered. He held his finger to his mouth. A hedgehog snuffled out from the long dead grass by the field edge. Oblivious to us, it trotted along like a Victorian woman raising her petticoat above her skinny legs, until it reached him, sniffed nonchalantly, then climbed over the toe of his boot, and on down the field edge until it was lost again in the grass. I was grinning from ear to ear. Grandad was beaming like a little boy, and said in a hushed pantomime voice, "That was Mrs. Tiggy-Winkle taking her washing home."

~

My grandfather's world, and thoughts, largely stopped at those walls, the edges of our kingdom. Beyond was someone else's concern. We had ties and obligations to those neighbors, and shared rules of decency. We had moments of collaboration, but what they did on their land beyond these boundaries was their business. Our walls, hedges, and fences were critical to our farming system. They let us manage parcels of land in many different ways. We had thirty or forty fields on the farm and lots of them were small. But each field to my grandfather had a character, almost a personality, and a backstory, which was part of a series of stories that made up a kind of epic poem. Those field poems came alive in the telling, and in the ongoing work. Knowing the fields and their whims and needs was vital: it determined what could or could not be grown in them. The yields some fields produced became part of their stories—some were of crops that were more bountiful than anything we had ever known—and we wondered if they were just myths.

Grandad said the sandier soil on the field he called "Castlebanks" was "hungrier" than other ground, and needed lots of well-rotted "hull muck" (the best kind of muck, from the barns where the bullocks were fed and bedded with straw, which had been rotting in a midden for months before being used) to enrich it. The "Bottom Banks" grew wonderful turnips, as big as footballs, which basked in the warmth of the sun. The "Eight-acre field" grew bountiful crops in dry years, but it was clay and could "sulk" in cold or wet weather and would cease to grow. Each field had a human history as well as its natural origins. He spoke of the men who had laid the hedge by the "Quarry field" or dug the drains on the "Railway field," or the various injuries men had endured in different places. He told of the two Buckle brothers, who had been fencing on "Merricks," and between each swing of the huge post hammer the brother who was holding the post would give it a shake with his hand on the top. His brother had been distracted by something and the hammer had come down and mashed the other's hand. I loved these stories. They made the fields the stage for a kind of magical drama.

~

By June the barley was up to my knees, and on windy days silver-green waves raced across the field. The field needed us less for a while, and I was swallowed up again by school. One morning as I was waiting by the church with the other boys and girls for the school bus, milling about, throwing stones and kicking pinecones, my grandfather walked toward us with his sheepdog and stick. He had taken some sheep to a field and was walking back to the farm. I knew he had seen me messing about. I didn't want him to think me a fool like the others, so I stepped away from the crowd. Perhaps I also stepped away from them because I sensed what was coming and didn't want them to laugh at him behind his back afterward. He stopped and asked what I was learning about at school. I told him we were supposed to be learning about the planets. He said he didn't know much about planets, he only knew about the sun. Then he told me about how its arc over the village changed with the passing of the year. He stabbed at the skyline with his stick where it rises on the shortest day in December, pointing to the southeast. "There—it comes up there," he said. Then he used his stick to create small loops over his head to show the path of the sun on the short winter days, explaining its changing arc through the seasons. And, in my eyes, he had turned into some giant insect as he moved through the movements of the sun across our land. One after another, he made the arcs with his stick above his head. He wanted me to know how the sun passes over, and that it was a thing of great wonder. He wanted this day to include at least one useful lesson, before I wasted the rest of it in school. He ended his little orbit on the fells to the northwest, where the sun sets. In my eyes my grandfather was heroic, but it was hard not to see that he was becoming an anachronism, even if I didn't have the word for it then: a man from another age, becoming irrelevant to most people.

The older kids were barely polite as they turned back to their schoolbags, not sure why this old fool, twenty feet away, was lecturing his grandson, and anyone else who would listen, about the rising of the sun, looking like some mad slow-motion samurai slicing the sky with a sword. And I was not sure either, but I loved his clock-like explanation of the heavens. He finished, patted his dog Ben, said he hadn't got time for chatting, grinned, and told me not to "misbehave and get locked up and miss the summer holidays," and walked away down the road, humming. The little red bus arrived. We clambered on and flung ourselves into the seats. A minute or two later we were down at the council houses, picking up those kids. I could see long gray lines of trucks on the motorway up the hill. A diesel train chugged down the tracks with black smoke rising from its exhaust, and through the finger-smudged window I watched the sun break through the clouds.

~

The teachers at our primary school were good and kind. They let one farm boy called Brian grow a crop of barley in his square of the school garden. He tended that crop as if his family's farming reputation depended on it. Every break-time he would be weeding or tending to it. And he stood beside it, proudly, when the teacher walked us past the different plots just before the summer holidays, past

the pathetic nibbled-off rows of lettuces, stitches of potatoes with blight that had withered and died, spindly carrots running to seed, and sad unloved plots where nothing
much grew except weeds. But Brian's plot stood waist-high,
swaying in the breeze, a perfect weed-free patch of silvering
green barley.

~

The six weeks of summer holidays seemed to last forever.
Every afternoon (when I wasn't missing in action, playing
with my friends) I would be sent to get the cows in for
milking. I'd take my bike, pedaling and swaying from side
to side on the steepest bits, until I reached the top of the
hill and could breathe and freewheel. This hill where the
cows grazed was called Burwens, a rough patch of unen-
closed common land, stretching alongside the road to the
next village. The cows were behind a temporary electric
fence—a thin wire held about three feet from the ground
by iron posts, with plastic insulated twirled heads, pushed
into the ground every twenty feet or so. The cows stood and
waited, tails swatting at flies, with swollen udders, bellowing angrily to be set free. I reached under the rusty electri-
cal box to stop the current running along the wires, my
hand trembling nervously near the flick switch. Its cabled
guts spilled out beneath it because it was so badly wired.
I had been shocked many times before. If I put my finger
half an inch to the side of the switch by mistake, I touched

a live wire and received a sharp electric shock. Once, when the switch was supposed to be turned to off, it wasn't, and as I grabbed the wire to unhook it and let the cows out of their pasture, I got a jolt of electricity. Occasionally a sheepdog would piss on the electric fence and earth the current, and yelp and run away as if it had been shot. The box was broken for years, and yet it never occurred to anyone then to buy another. The unspoken ethos of the men was always to manage with half-broken things or mend them. We called it "making do."

Burwens was covered in tussocks of long grass, wildflowers—white, pink, and yellow—purple-headed burr thistles, and ragged thickets of gorse. It was the kind of half-used, half-wild space that existed then in many landscapes. My grandfather paid rent simultaneously to two different landlords for it. The community elders believed it was "parish" land and always had been—it was theirs collectively. But the local aristocratic estate that owned the farm said it was their land; they said they had registered it legally. I had no idea who was right or wrong, but my grandfather, always keen to avoid a fuss, said it was simpler to respect both sides and paid the token rent twice to two different owners. When the cows had grazed down one part of the common, my grandfather would lift out the iron fence posts and move the paddock to another ungrazed patch to let the first area recover. The new area was often as deep as the cows' knees. The whole common was at different stages of regrowth, a quilt of different greens. The land the cows were currently on was hard-grazed to a lighter shade of green and dotted with cow muck. Only the inedible gorse and the bitter-tasting ragwort, plus a few thistles and nettles, stood above the trampled ground.

~

That summer my grandfather taught me to hate ragwort. He, like all farmers, despised the stuff. He said it was "toxic" to grazing animals and could kill cattle. Its yellow flowers were symbolic to him of a kind of neglect. Land that was covered in ragwort was being "let go" through bad farming. He enlisted me and my cousin to "knock back" the ragwort (every grand plan for me occurred on a Saturday). He told us he would pay 10p for every plant we pulled clean out from the ground. We must not leave the roots, he said, because they would grow again. The ripped-out plants were to be thrown in a heap to burn.

We marched up the hill and pulled the plants out, until there wasn't a single standing yellow flower on the common. We were proud of working so hard and felt we had exceeded my grandfather's expectations of what we were capable of. He had thought we would make our excuses and slope off after an hour or so, but we had kept going until we were done. Our hands stung, were stained green, and stank of the plant for days afterward. We didn't get the hundreds of pounds we were sure we must be owed, but five pounds to share between us. Grandad then went and forked the piles of wilting flowers into the trailer to be burned on a bonfire. And I was told to get the cows home.

~

The cows strode home down the windy road. Behind them rose a cloud of black flies. The cows swiped their tails and tried to squash them with their noses into their flanks. As the cows walked, they splattered jets of grassy-green shit that formed swirling trails on the tarmac. The muck baked on the road like pancakes, peeling and pulling off, as it dried, the tar-skin from the road. I steered my bike between the cow shit, balancing on the pedals to go slow enough behind the older cows. Swallows dipped in and out of the leafy lane below the breeze. The verges were three feet deep in white flowering cow parsley. My grandfather called it "kesh." Its lush aniseed smell filled the hollows. Halfway down the hill was a little wood where some of my friends were building a den. I was tempted to go and play there, but I couldn't leave the cows. We were always building dens in messy abandoned bits of the village, including the rubbish dump, where people threw their old tires, TVs, and mattresses. We harvested crab apples from the hedgerows for "food" and stored them. Whenever we dared to bite into one, they were so tart they instantly turned our mouths to fuzz.

~

At the bottom of the hill, the cows were meant to walk through the gate to a collecting yard, but they had escaped and headed up to the village green. My father, who was back in the yard, saw them and came to help me. He shouted that they were "just taking the piss" and ran past them, shouting and wielding a length of plastic pipe to whack them with. His dog Lassie joined in, nipping an ankle or two and barking at them until they were all in the yard.

In subsequent years the village became more middle-class and tidy, and cows plowing up sections of the village green with their hooves, and dirtying roads, became a source of tension with other villagers, but when I was young, everyone knew that cows mashed up the verges in wet weather and splattered the road with shit—and they always had. There were water troughs in our yard, and until a few years before, anyone with cattle in the village had been free to walk them into our yard for water.

~

My father was as tough as old boots, but he loved his cows. Until a year or two before, at the beginning of the 1980s, he had a herd of eighty black-and-white dairy cows—Friesians, with broad fat backs and sturdy legs. They grazed outside on the pastures for about half the year, from May through to October. The other half of the year, from November to April, they lived in the byres (cow stables) and barns and were fed hay. But Dad had to sell his beloved dairy cows to raise the money to buy the land that had come up for sale right in the middle of our fell farm (the very land my family now lives on, in a barn that was then home to only owls and cobwebs). Land in the midst of your own farm was so rarely put up for sale that he felt he had to buy it somehow, despite having no money.

Even so, having drunk milk only from his own cows his whole life, he refused to have shop-bought milk, which he judged to be a pitiful thing. He said it was "watery" and "messed about with." He called skimmed or semi-skimmed milk "pigeon milk." So he kept this handful of the older cows that weren't worth much for milk "for the house," and for rearing calves. The milking was done with a little electric machine. The milk came out of the "unit" (basically a milk churn) thick, frothy, warm, and creamy. The inside of this unit quickly became coated with a yellowy fat from the milk. You could never scrub it often enough to keep it clean. The flies that followed the cows into the byre from the surrounding fields often ended up in the milk and had to be lifted out, legs wriggling in the yellow froth. If the electric went off, we would milk by hand into a bucket. I loved firing the milk from the teat into the foamy warm milk in the bucket, loved the soft frothy drilling noise it made.

Dad's favorite was a black Friesian with a friendly face and big, black, glassy eyes that sparkled as she chewed her cud. He called her "Old Blackie" and insisted she was "about twenty years old"—but she was always about "twenty years old" no matter how many years passed after that. She was a gentle cow to milk, "letting her milk down" and never kicking whoever was milking. Out of love for Old Blackie he also kept on her best friend, "Snowy." But she was a "complete bitch." Once, she knocked me over in a doorway from the byre, and only by luck did her hooves not crush my rib cage. When I got up, my father's face was ashen, and he said he'd have to sell her, or "maybe I'd have to learn to get out of the way faster." She was wild-eyed, swishing her tail to thwack me, or lashing out to kick if the flies tormented her flanks. He said I had to "boss her." He would lean into her flank, with his whole weight so she couldn't kick, until he was almost lifting her half off the ground, and then reach under her confidently and attach the pulsating suckers onto her teats. When she calved each year, Snowy was even more dangerous. She would bellow at anyone who so much as looked at her calf in the first day or two. I was told to stay clear. A few years earlier, in a maternal rage, she had killed a farm cat that had dared to walk past her newborn calf. It lay squashed on the concrete yard, its eyes bulging out of its skull. Why he kept Snowy was hard to fathom, but my dad was always a little bit random.

~

With hindsight, I can see that our farm was full of animals and places that defied my father's and grandfather's best efforts to tame them: the stackyard full of old machinery, chest-deep in nettles; forests of tangled thorns, like those that grew around the castles in fairy tales, in the abandoned quarry by the road, where bullfinches sang from the thickets, their barrel chests a bright plum color; rotting tree trunks at the top of the yard that had never been ripped out, and were now crumbling and full of red ant nests; the edges of the common land where the cows grazed, scruffy and half-wild; and even the fields of barley and oats were speckled with poppies and weeds, the pastures were full of thistles, and the hay meadows, by late June, were full of wildflowers.

~

My grandfather marched into the "Long Meadow." He gestured for me to follow in his wake of shadow and shaken dew. We were in the lull before the grass was mown for hay. At breakfast my father had said we should mow now, and "get on with it" like our neighbors. Grandad said he didn't think it was a good idea, but he would go and have a look "with the boy." At the field his tweed jacket was slung across the wall-top. He was stripped down to a sweat-stained undershirt, arms leathery brown to the elbow, then a milky white where his stripy cotton shirt had always been (at school, the town kids called this a "farmer tan"). He removed his cap, and his hair flopped away from his pale scalp. He walked on into the greenness, and then, like an old heron half-bent at the middle, he picked at something, again and again, from the meadow. I knew I was to be taught something.

After a minute or two, he came back with the grasses splayed like a fan in his hand. It was time I knew grasses by name, he said. So he pointed at each one and listed their names. It was something I was meant to take to heart. Meadow Fescue. Timothy. Common Bent. Cock's Foot. Yorkshire Fog. Ryegrass. Rough Meadow Grass. Sweet Vernal. Meadow Foxtail. He said that each species of plant told a good farmer something. The good grasses and plants showed that the soil was full of fertility and "management." The worst ones, the "weeds," told him that the field had been losing condition, that they had been robbing it of nutrients, taking more than they had put back. The Fescue flopped over his rough old hands, its head drunk-heavy with tiny oat-like seeds. Look at this, his face said, pay attention, learn this, a farmer needs to know these things. But I was daydreaming about the clouds above, not paying attention, and forgot all the grass names for many years.

As we walked back to the house, he mumbled that the hay was nowhere near ready; he would not mow it now, just because my father had no patience. We would mow it later, as we always did. Back at home, Dad said that Grandad was living in the past, that we hadn't got the manpower for doing things the way they used to, and that silage was better feed for the cows. He said we were farming like it was still the "fucking 1950s."

~

I was scrunched up in the back of the Land Rover with the sheepdogs dripping candles of saliva onto my legs. It was a Saturday and we were traveling to work on my grandfather's farm, to worm the lambs. Dad and John were in the front talking about football, but also passing brief comments on all the farms and farming families we passed.

"That's a tidy farm."

"They are tremendous stock people."

"He was caught in bed with another fella."

"He never stops bidding when he wants a tup."

"She's lovely."

"They own half the village."

"Best cows in North of England them."

"He has no work in him."

"That's a total shithole."

"Smartarse. He knows the fucking lot."

I was wedged between burlap bags of sheep feed, fence posts, rolls of fencing wire, buckets of hammers and nails, bits of railing, and tubs of sheep-marking. It took half an hour to travel between our two farms. Until that summer, the journey had been a blur of green flashing past, and none of it meant anything, but as my farming education progressed, I began to see what those fields were. I didn't know it then, but my life was a mirror of that journey: a movement into the fells, a journey the wrong way, backward away from modernity, into one of the last traditional farming landscapes.

The first half of the drive took us through the lower ground by our local town, where the land improved and the farms grew bigger, the farming methods changed to fast-growing herds of modern dairy cattle or giant fields of barley or grain. I could tell from the men's talk that these farms were more prosperous and the farmers wealthier than us—we half-admired, and half-resented, them. I knew already that this was good land, with flat, deeper soil. I could see the big tractors and shiny machinery in the barns and the fields. I could see the giant barns and cattle sheds being erected. I sensed from the talk in the front of the pickup that these valley-bottom farms were changing to something else. I didn't understand all that they said, but whatever was happening to these farms was coming for us too. My dad seemed nervous about it all, as if we were falling further and further behind these farmers who were leading the race. We were never going to have the money for all this.

Then we turned left at the roundabout on the edge of town, with its gray industrial estates, giant feed mills, chicken-guts processing plant, and motorway junction, and headed for the Lakeland fells to the west. As the road wound uphill, we passed out of the best land and eventually found ourselves on the little road into the valley where my grandfather farmed, nestled between the two rounded fells that marked the start of the Lake District. We wove be-

tween small meadows, bounded by shaggy hedges and silver walls, and rough grazed fields that rose from the boggy
valley bottoms to woodland and wilder fellside. These fields
were like a time capsule, farmed in the traditional way.

~

As I spent more and more time with my grandfather, I became aware that his fell farm was barely "improved" at all.
It was staggeringly beautiful, even the most hardened and
unsentimental workingmen we knew would comment on
it. That outdated farm seemed to cast a spell on everyone.
The problem for my father was that it was full of work that
had previously been carried out by a small army of skilled
men and women, and they were disappearing, leaving him
rushed off his feet trying to hold the two farms together as
my grandfather grew old. Across the two farms, he would
be shearing sheep until it got dark, then milking early the
next morning, then racing off to dose some bullocks on
some distant rented land, then wolfing down his dinner,
then spraying a field of barley, then trying to get back to
milk the cows, then mending a fence from which some
sheep had escaped. It was exhausting and I'm not sure my
dad understood clearly what had changed, or how impossible his fight to keep up was. John was also getting older
and needed lighter work. Within a few months he went to
work for a local builder doing odd jobs—and my mother
and I were enlisted to help plug the gaps.

~

Thirty years earlier, half a dozen men would have gone with scythes, and with their shirtsleeves rolled up would slay the thistles each June or July. Now there was just Grandad with his scythe, Dad on a tractor, and me with a sickle. Sheep and cattle don't eat thistles, so grazing beneath them is almost impossible if they grow strong and dense. Thistles were one of nature's ways of taking our land back, making farming hopeless. They ruined the pastures.

The fields that were too stony and hilly to be cropped, and with large banks or edges too steep for tractors, were used as permanent pasture, and it was here that the sea of thistles was chest-deep. So my grandfather and I had the job of slashing them down by hand while the tractor cleared the civilized bits of the field.

~

He stopped every few minutes to catch his breath. As we worked, swallows and swifts tacked backward and forward after the insects we disturbed. They passed on either side of me like star fighters and snapped their wings so close I could almost feel them on my cheek.

My grandfather spat on the blue-gray whetstone, then pushed it through the arc of the scythe's blade. Its gritty surface rasped at the cutting edge, leaving it shiny, metal-raw, and sharp. He pulled his finger gently along the blade, making me wince. He smirked at my nervousness. When he was satisfied it was carving-knife sharp, he stuffed the sharpening stone back in his pocket and put his scythe back to work. Soon the handle curled and twisted around his body in the classic style. He swung it effortlessly, curving around him, again and again, so that each slash took about a six-inch swathe of thistles. Occasionally, he cursed as the arced blade hit a hidden branch and the scythe was wrenched out of rhythm. His shirt was specked with the thorns of thistles, and the scythe blade itself was wet-green with thistle sap. Twenty feet away a goldfinch swayed gently on the purple flower of a burr thistle and rocked back and forth, its little gold wing-bars flashing in the sunshine as it plucked at the thistledown. I swung my sickle, or "slasher," back and forth. Slash. Slash. Slash. Thistles were flung past my shoulders on both sides. My neck was wet with sweat and itchy with thistledown. My arms and neck were sunburnt brown. My half-length hair was bleached blond like the kids hanging out on the beach in the movie *Jaws*. The wooden shaft of the sickle had blistered my thumb. If I caught too many thistles with a swing, it snagged, stalled, and twisted my wrist painfully. Across the field, my father was driving the tractor with the "pasture topper" on the back. The two swirling blades vibrated and hummed underneath it, as they slashed everything off at about three inches from the ground. Occasionally the machine hit a rock and there was a crunching metal sound, and cursing from the tractor window.

My grandfather told tales of how poor this land was when he had started farming it in the 1940s—just thistles, broken fences, and unpicked stones. Decades of his work meant it was now "in good heart." But, in truth, it wasn't a battle won. A month later, many of the mown thistles had regrown and seeded again.

~

In August, the barley began to change color, slowly ripening to gold. The heads, whiskered and silver, turned back to the earth as they dried. My grandfather would rub the ears of barley between his palms, until just the grain remained. Then he would pinch one to his mouth and bite it. He told me to copy him. Until they were ripe they were full of squidgy milky-white fluid, but as the days passed and the sun did its work, they became hard, until his teeth split the kernels like a chisel. Then he gently spat the two halves onto his hand to reveal the densely packed flour within the grain. That told him the crop was ready for harvesting. And for the first time in my life I felt a kind of pride about a crop, because I had helped to grow it and knew how much work and faith had been invested to get this far. A few days later, as our school bus passed down the little road, I knew what the wild black specks above the barley field were too—and my heart sank.

~

My father was striding toward the field, down along the hedges, past trees with leaves covered in a silver-gray coating of dust. Soon the field ahead was echoing with cawing, and the sky above it was a swirling plague of crows. The crows had been ignoring the scarecrow, dressed in Dad's wedding suit and stuffed with straw and tied at the wrists with bale string. They had also been ignoring the mechanical crow-scaring device he had put in the field for several days. They were too clever to fall for such tricks. They were ravenous, smashing great holes in the crop with their wings to steal the ripening corn. Dad meant to shoot a couple and hang them from makeshift gibbets in the field, to frighten away the others and save the crop from destruction.

He had his 12-bore shotgun slung over his shoulder, cocked in half. Cartridges stuffed in his pocket. He gestured for me to be quiet and walk quietly in his wake through the waist-high barley. He cursed a handful of crows that saw us. They headed off, cawing a warning. He made his way quietly down the field, hidden from the crows by the hedge and the shoulder of the hill. A babbling, raucous clatter of harsh calls told him they were feasting not far ahead. Then, as we got within sight of them, they took to the sky, turning it black. Hundreds of rooks, carrion crows, and jackdaws crowded the skies in an orgy of blackness. They flew every which way in a panic. Every which way, that is, but within a hundred yards of us. I was crouching behind him in the barley, which hid us. The barrel of his gun was poking up against the sky. He said this was a total waste of time, that the crows would circle the parish, land in some of the ash trees at the other end of the field, and mock us until we had gone. Then they would resume their wicked feast. But some of the rooks were so brazen that they were returning, testing the skies for threat. They didn't really know where we were. Dad remained rigidly still. The air was tense. Silence, except for the sound of their wings. Two or three black dots approached high above. The first one was much too high, impossible to kill. But Dad leaned back until he was straining, almost crushing my legs. A black speck, wings outstretched, moving slowly, but so high it seemed tiny. He pulled the trigger gently. The gun recoil sent a shudder through him to me. The shot had gone, but he was still peering up the little sight on top of the gun. The air smelled of cordite. Then, high above him, the bird crumpled into something smaller and fell from the sky. It landed about five feet from where we were crouching on the bone-hard ground with a feathery thud. All hell had broken loose. The crows knew that we were there now and had a shotgun. They fled like a kind of storm wind that sucked the air from the field.

Dad stood up and looked at the terrible damage they had inflicted on the barley. He took in a sad breath. Two acres or more of it had been flattened. It lay with broken stalks, as if a herd of elephants had rolled on it. Everywhere was flecked with white crow shit. Black feathers lay amid the barley. I felt proud that my dad was such a good shot. We left the dead rook hanging from the lowest branch of an ash tree, bobbing by a leg from a length of red bale string.

~

The "combine man" arrived later that August and irritated my father by going on and on about the destruction the crows had made. Lice-riddled, they cawed at us from the distant ash trees. Like most small farms, we couldn't afford a combine of our own, so a local machinery contractor sent a man to do the harvest. The giant red machine rumbled down the lanes, chugging out clouds of black exhaust fumes and pushing back the branches. Behind it followed its own reel, a cylinder rake on spindly trailer wheels that clattered like a shopping trolley with every bump in the road. The frustrated, slightly grumpy, and bewhiskered driver was the younger son of one of the old established local farming families. My father called me from the stubble to ride up on the combine and watch it work. I squeezed in next to the driver, the handbrake jutting awkwardly into my backside. As we drove down the field, the grain was raked by a giant revolving reel onto dozens of little serrated triangular knives that severed it from its roots, about four or five inches from the ground, then it was rolled by the windmill-like cylinder into the heart of the machine. The stubble left behind the machine had a neat fresh-shorn look. Waist-high rows of straw dropped from the back end of the combine and divided the field into rows. Behind the driver's seat was a giant tank to hold the grain, and through a glass panel I could see it, cut from its stems and shaken from its ears, pouring out.

The combine circled the ever-diminishing patch of barley, trailing a cloud of dust. At forty-five-minute intervals, my father returned to the field on the tractor, pulling the grain trailer alongside the combine. The combine extended out its giant red arm to empty its belly through the arm into the trailer. The ground was hard, and we jolted around on the rougher headlands. The combine man scratched his beard from time to time and loosened his neck scarf to shake off the dust. He had bags under his eyes and a packet of cigarettes stuffed in his top pocket. Ladybirds crawled along his arms. Every surface on the combine was covered three inches deep in dust and chaff. As we cleared the field, diminishing the standing crop strip by strip, rabbits leaped away to the dikes, and the driver's dog, running down below, hunted them across the stubble.

The combine driver cursed at the poppies and thistles, and told me that my dad should have sprayed them with pesticides as the crop wasn't as "clean" as on the more modern farms he had worked at. He said my grandfather was a bit behind the times. Weeds could be eradicated now. There was no excuse for them anymore, he said. He told me his friend drove a much bigger combine, but it would be no good in our small fields. We needed to widen the gateways and clear some of the shaggy old thorn hedgerows to make bigger fields. Small fields like this, of odd shapes and elevations, were hopeless. Trees drove him mad with their branches that scratched the paint from his machine. Everything was wrong with our farm. I was silently angry, but I held my tongue. When I told my grandfather later on, he said all this "modern talk" was bullshit. If the combine man was so smart, why was he just driving a combine? He said we needed to produce as much as we could of what we needed on our own land, and we needed the barley and the straw for our animals. The stubble left behind was somewhere to spread the muck in the depths of winter, returning the goodness from the animals to the land.

I rode home with my father down the lane, three or four tons of barley in the trailer behind us. The brakes on the old tractor weren't good, so he pumped them nervously as we went down the steepest bit of the hill to stop it "running away with us." Back at the barn, the barley flowed out of a sluice gate in the back of the trailer and down to the concrete yard.

My father tipped the trailer at the bottom of an electronic auger (a motor-driven Archimedes' screw inside an aluminium pipe that takes the grain up and into the barn). Then he stood up to his knees shoveling grain to the auger. He sank into it as if it were quicksand, so he had to lift his feet out every few seconds, treading back to the surface. Feet weren't meant to do this. He was sweating. In front of him the grains of barley were being sucked away as if by some invisible force, a whirlpool pulling him and everything else down, down, beneath the surface. Tons of grain, strands of straw, and ladybirds were swallowed and churned up the tube and spat out up in the loft. It looked as if he was standing in a giant egg timer as the sands ran away beneath his feet. It was evil, strength-sapping work. The machine went on and on, and vast sweeping shovel-fuls simply bought him a few seconds: it was hungry and it demanded more. I was told to stay away from it. Stay where I was in the back of the tractor. Never go near the auger, he said, it was dangerous. It had a cruel power that didn't respect flesh and blood. A farmer we knew had been sucked into a machine like this and lost both of his feet—flesh torn away to stumps. But he got out of hospital quickly and, with newly fitted synthetic feet, resumed working like an able-bodied man on his farm.

When the last load was led home, the stubble shone in the afternoon sunshine. A few days later and the straw was baled and put into the loft in the barn, directly above the byre where the cattle lived in winter.

To my boyish eyes the work of our barley field was done, the cycle complete. But my grandfather thought no such thing. Growing and harvesting were merely a prelude to the dark months when his days were full of feeding and bedding livestock with barley and straw. The straw would be thrown down through a hole in the loft floor for the cows' winter bedding.

At some point that autumn I became aware that no one was telling me to go out and work now: I was choosing to.

~

Harvest festival. We trooped to the church from school and sang hymns like "We Plough the Fields and Scatter." It was one of the highlights of our year. There was an embroidered banner in the church of a young Jesus, who looked very blond and Anglo-Saxon, and around his feet were little birds eating the grain he had thrown down for them. That night there was an auction in the village hall. Families walked from their farms and houses, in the dark, with torches. Everyone crowded in the little back room. The women poured tea and handed out custard creams and pink wafer biscuits. The farmers talked about the weather and whether they had enough crop for the winter, or how the sheep sales were going. They gossiped about the crop wasted by the farmer down the road, as he had missed the dry weather, but he wasn't there to hear it. The farmer they were talking about was a newcomer who had been to agricultural college, and they were delighted by his incompetence.

The kids raced around the large room next door, playing tag. There was a tapestry on the wall that the children at the village school had made ten years earlier, with all the farms on it and the names of the fields. At the back, the vicar's wife was organizing all the items for sale. The trestle table groaned with food the women had baked, or given, ranging from beautiful golden loaves of bread in the shape of a sheaf of barley, to bags of homemade fudge that

we bid eagerly for, to jars of freshly made jam and mar-
malade, to tins from the back of cupboards of pineapple
chunks and canned soup that no one really wanted. We
flapped our hands around in the auction and bought what
we could. My dad was the auctioneer for the night. When
it was done, the vicar said he wanted us all to go to his new
Sunday school—but none of us were listening because we
were stuffing our faces with my grandma's gingerbread.

~

Grandma's kitchen in the farmhouse became a jam factory every autumn. She didn't reckon much of "men" in her
kitchen, but I was still a "boy." And, on this particular day,
I was not gainfully employed so was recruited to help and
learn. Grandad drove us to a lane a couple of miles along
the valley. It was edged with scraggy bramble hedgerows,
and was Grandma's favorite gathering place. She wore Wellington boots that met her brown skirt at her knees, with a
thick quilted jacket and headscarf tied under her chin. She
had thick bottle-top glasses for her short-sightedness. I was
to help fill her Tupperware tubs with fruit. Grandad left us
and said he would come back later; he had to go and see his
cousin "about some sheep wintering." We walked up a "lonnin" (lane) for about half a mile. It was an old way for sheep
and cattle to get to and from the fells from the farmsteads
in the valley. As we climbed, the ash trees with their thin-
ning leaves were left behind. The fell above us was cold,

bracken-brown, autumnal and bare, but before us the fields were a deep dark green, and we could see the valley stretch for miles. The air smelled of sheep "purl dip," from ewe lambs that had been made ready for the sales.

The sides of the lane were rife with thorns and brambles and soon we reached the place she had come for, and she set down her containers. The tangled mass of thorny vines was black with ripe brambles, and we picked them, arms scratched chalky with the barbs. I stuffed every third or fourth bramble in my mouth, while Grandma told stories about when she had picked fruit with her "momma." When we walked back down that lane, we were laden with fruit. My grandfather was waiting, asleep in his car, the windows steamed up because of the heaters on inside.

Back home, Grandma scurried around, adding this and that, and stirring, keeping a watchful eye on the thermometer, until it became a thick, puckering lava in her brass jam pans, and the house was heady with the smell of bubbling fruit. She had already been on other expeditions in the preceding days with her sister and cousin to a "pick-your-own" at a fruit farm some miles away, returning with baskets of strawberries, raspberries, redcurrants, and plums. From her freezer, she brought out other fruits she had foraged for, or grown in her garden—gooseberries from a hedgerow, rhubarb from the garden. She waged a noisy, pan-banging war with the blackbirds and thrushes to protect her fruit bushes. And what didn't make jam was turned into pasties on a metal plate.

Apples, pears, and plums from the orchard were stored for winter under the spare bed. They were laid out, never touching, on opened-out sheets of the newspaper. Grandma would climb stiffly down beside the bed, pull out the sheets, turning the fruit carefully in her hands in case of rot or mold, and those that were turning bad would be cooked. When we came in from the freezing cold in winter, we would be met with plates brimming with hot food, followed by a Queen's Pudding or an apple pie and custard.

My grandmother was an expert at turning the things the farm grew, harvested, and reared into meals. Almost everything she cooked was home-grown, seasonal, and local. Meat, potatoes, vegetables, or fruit and berries were made into preserves and chutneys. There was the occasional orange or banana that had been bought, and a few packets of crisps were kept in the "drinks cupboard," but otherwise plain traditional food was placed on the table to be eaten and enjoyed. Grandma never asked what we wanted. There wasn't a choice.

She never bought or cooked (or trusted) food made in factories by strangers. She cooked to a schedule that seemed to have been handed down by God to Moses (or Mrs. Moses), it was so fixed. The main meal of the week was the Sunday dinner—a well-cooked lump of beef usually, but often a leg of lamb or a shoulder of pork. Any sign of blood or pinkness in the meat was considered dangerously continental. The meat had to be properly cooked until it was dried out, and potatoes were roasted. Every meal featured potatoes: they were peeled on a Monday, plopped into a bowl of water for the whole week, and brought out to be mashed, boiled, chipped, sautéed, fried, or roasted. The "leftovers" were cooked again. Cold beef was eaten in sandwiches for the rest of the week. We killed our own sheep and cattle in the byre, bled them into buckets hanging from the beams, skinned them, and cut them up on the kitchen table with a saw and a set of knives. Every bit was used, not just the best joints, but the oxtail and the tongue and the offal bits; even the blood was used for black pudding.

Grandma had home deliveries from the local bakery on a Thursday night, and the driver came in for a cup of tea and a slice of gingerbread, and to swap racing tips with my grandfather. On a Tuesday, Bert the Butcher came and sold her a few slices of cooked cold ham, and maybe some sausages. On the whole she didn't rate shop-bought food—that was just throwing money away. I can only remember going to one restaurant with my grandparents, on some anniversary or other. My father had a row with the snooty waiter because he called the turnip he served a "swede" and Grandad said he was daft, because it was definitely a turnip.

At times Grandma did her housework with an intensity that suggested something else was on her mind. I would later learn that she had once tried to leave my grandfather for fathering a child with another woman. She had attempted to return to her childhood home, but her father had closed the door on her and told her to go back to her husband. You made your bed and you must lie in it, they said.

~

Mum's life was different from Grandma's. She worked hard in the byres, feeding cows and mucking them out, doing the work of the hired men who had worked on the farm until recently. Then she would have to make a meal and present it to the table about ten minutes after kicking her farm boots off. She seemed to rather like the farmwork, but to resent the endless drudgery of the housework. My grandmother thought it was all wrong that my mother couldn't be the perfect housewife because she was outside doing farmwork. The men had failed if a woman couldn't be left alone in the house to do what she was meant to do.

~

My parents were half-broke. I could see it in the second-hand tractors, rusting barn roofs, and old machinery that was always breaking down and never got replaced. But I could taste it too, in the endless boiled stew and mince that was served up. I ought to have been grateful for this whole-some home-reared beef or lamb, served with mashed pota-toes and peas and carrots, but I wasn't. I hated it. I spent many mealtimes standing on the porch chewing meat to a cardboard pulp, sent there by my dad for being "ungrateful" and not "eating the food set in front of me." Sometimes I would feed the sheepdogs my food and lie about having eaten it. Other times I would just stand there daydreaming with a mouth full of over-chewed meat.

A gray air-raid siren stood in the corner of the porch. My father was supposed to take it to the top of the hill and use it to warn everyone in the parish in the event of a nuclear holocaust. Mum was supposed to fill the bathtub with fresh water and seal the windows with tape and newspapers. After mealtimes my father would look at me still standing there chewing "perfectly good meat" with a look of confusion and disgust. He'd stride off up the yard, and my mother would come out and say, "Quick, get a cheese sandwich and eat it before he comes back and kicks your backside."

One frosty autumn day my father came back to the house after dinner, crestfallen, because Old Blackie had "gone off her feet." We looked where he pointed at the top of the Cow Pasture, and saw her slumped on the ground. She was worn out and had reached the end of her days. Snowy was grazing sympathetically nearby. Dad knew that she must now be shot, dragged down to the yard, and sent away with the "knacker man." He took his shotgun and trudged sadly up the field. From the kitchen we saw his head sunk in sadness as he had a word or two with the old cow. He put the gun a few inches from her head. And then suddenly Old Blackie rocked forward to her feet, pushed past him and his gun, and wandered off to graze with Snowy. As he passed us in the kitchen, on his way to put his gun back in its cabinet, we were all grinning. He told us to "shut up," but he was smiling. She lived quite healthily for another eighteen months.

~

When the ground became sodden, the cattle were brought down from the fields for the winter to live in the barns and byres, and it was my job to go out and help feed them when I got home from school. The dusky sky echoed with rooks and starlings swirling around the village on their way to roost as I tramped across the yard to the "meal house" where the cattle feed was stored. It was deafeningly loud there because the barley crusher was on, its two giant flywheels flattening the thin flow of barley. We filled burlap sacks with it for the cattle and poured it across their hay and silage like a scattering of pale yellow corn flakes. I had to brush up the sweepings of waste barley from the meal-house floor for the hens.

I was less reluctant to do these chores than I had been the previous winter, but I still hated the henhouse. It was a small wooden shed, with an outside wire-mesh run full of nettles. The hens scratched and pecked in the dirt each day in the sunshine and caught the occasional juicy worm or a spider, and retreated to the shed at dusk. There was no electric light in the hut, so as I stepped in, the door swung shut, trapping out the last of the daylight. It was dark now except for the small rectangular light of the pop-hole through which the hens went to and fro from the outside run. The floor of the shed was raised about a foot or eighteen inches with years of dried hen muck. Deep in the shed, the laying hens sat silently in the nest boxes, except for an occasional cluck or wing flap in the darkness. They were motionless until my groping hand found them. Then there was a minor explosion of clucking, feathers flying, wings flapping, and dust swirling. One or two came squawking through the pop-hole as if a fox had entered their coop. But the "broody" ones held tight to their nests. I reached under their feathers into their warm downy underbelly to take the eggs from beneath them. The older hens stabbed at my hands with their sharp beaks, drawing blood.

But then something else moved in the dark. Rats. Big rats, fat from eating the hen feed. I could sense shapes moving on the floor, the darkness electric-charged with fear. Suddenly I heard squawking and saw a huge rat trying to pull a half-bald hen down a hole in the floorboards. I shouted and the rat let go and vanished. I retreated from the hut, shaken, and told my father, who said, "Them rats want fucking poisoning. This is past a joke." And he must have done it, because for the next week, whenever I collected the eggs, I found dying or dead rats in the pen, being pecked at by the hens.

The reward for this ordeal was a handful of fresh eggs each day. Back in the warm light of the farmhouse they were a range of browns—some speckled and odd ones smudged with hen shit. They warmed my hands and I liked that. The eggs had a deep orange yoke when they were boiled for breakfast. And to keep them clean, I carefully filled each nest box with golden barley straw from the barn.

~

That winter I began to see, for the first time, the cruel pressure that was weighing on my father. Long after I was tucked up in bed, he would be working out in the night, hammering away in the workshop or welding some broken hayrack. Or I'd hear his voice loud and angry through the floorboards. I heard the other kids at school talking about where they were going on their winter holidays, but my family never had a holiday. And my mum was caught somewhere between the kitchen sink and doing the work once done by a man on the farm.

~

Mum was shaping the dough on the table in our little kitchen. It stuck to the tabletop, despite the flour she had sprinkled on it. She had a turtleneck on. The sleeves were pulled back to her elbows, exposing her long pale arms, which were pushing and pulling. She flicked her hair behind her ear. She was slim and pretty. She had tried to follow the recipe in the Be-Ro recipe book. My grandmother had told her it was easy to make scones, but my mother found such things almost impossible. Maybe Mum wasn't cut out for being this kind of farmer's wife. Smoke was rising from the oven, and a strange noise emerged from my mother like a stifled scream. She opened the oven door, pulled out the tray, and stared at the carbonized lumps. She looked as if she was going to cry. She mumbled to herself and said she didn't have time for this shit. She had her outside jobs to do, and the clock was ticking.

~

Dad splashed on diesel and then flicked a match at it. WHOOSH. The giant heap of thorns burst into flames like a disaster at a chemical plant. Soon, vast clouds of black smoke rose from the orange glow of the fire to the darkness above. Dad and John had built the bonfire on the common land from farm rubbish, plastic bags, tree roots, thorn branches from the laid dikes, old furniture, stinking road-bald rubber tires, bags of bale string and plastic jerry cans of waste oil. Anyone else with stuff to dispose of would also have dumped it on there over the past few days. Occasionally an aerosol exploded in the heart of the fire. Everyone ran backward as if a small bomb had gone off. The men casually lit fireworks with cigarettes and shoved them into empty lemonade bottles for launch pads seconds before they zoomed into the sky. Mothers shouted at the children to stand farther back—from the flames and the rocketry. Grandmothers plied us children with baked potatoes, sausage rolls, and treacle toffee like black shards of glass that threatened to break our jaws and rip our teeth out. Then, an hour later, we all strode home in the dark, stumbling over branches or clumps of grass. Some kid was wailing, yanked away home by a frustrated parent because they had burned their hand on a sparkler. Flashlights shone into the trees and up above to an empire of stars. Frost nipped at our faces. We passed a fox on the way home, and my father called it a "thieving bastard" because, he said, one of its brethren had killed some of our hens a year earlier.

~

I escaped to my grandfather's farm in the Christmas holidays. The old man was getting weaker and needed my help to feed his suckler cows now that they were "laid in." Plumes of steam rose from the byre door. A fog of familiar warm smells drifted across his cobbled yard: cows, hay, shit, and piss. The cow byre had stone and lime mortar walls, a sloping green-slate roof, and a wooden door in two halves like a stable, with an old iron latch. Cobwebs hung from the rafters like tangled pairs of women's tights. Nineteen cows lived up one side of the byre in winter, tied up by their necks in their stalls. In front of each one was a stone trough for hay and a cast-iron drinker that they operated by pressing down with their mouth or tongue. Sunlight shone down through holes in the slates. And in these celestial beams the hay dust danced.

I helped my grandfather carry hay bales from the shadowy barn at the far end. An owl ducked out through a broken window into the daylight, a blur of movement that never quite seemed to be a whole owl. Grandad didn't like to disturb it, so he got his hay quietly, whispering, almost with veneration, as if the owl owned the barn and we were trespassing. Big fat brown moths fluttered about when we moved the bales, and often ragged-winged butterflies as well, like tattered jewels. Back in the byre, a robin followed my grandfather on his rounds, waiting for the little handfuls of hayseeds that fell from his jacket pocket whenever he fumbled for his penknife.

The cows bellowed at him to be fed, and he talked to them, telling them to be patient. He threw each cow a slice of hay. They flung the slices around in front of themselves with their mouths, ripping off giant mouthfuls with their curled tongues. The whole byre echoed with the chomping. Their neck chains jangled and clanked with the throwing up of their heads. I was not as brave at walking between the cows with slices of hay as my grandfather. Some of them lifted a leg threateningly, poised as if about to kick. I backed off, scared, but Grandad knew them better. He passed by the cow's flanks twice a day. He walked through clouds of their breath, was licked or nudged by their mouths, and occasionally dodged a kick.

He watched their tails for discharges of menstrual

fluids, which revealed whether they were in calf, as intended, or coming "a bulling" (coming into season). The cow could then be walked to the bull in another barn to get mated. When a cow was due to give birth, he would go out through the night with his torch, me with him, to check on her. His breath was caught in the torch's glow, for a brief moment, before it rose to the rafters. He peered into the loose boxes, or "hulls," deep in golden straw, where he put the cows to give birth. If the birthing calf's legs were large, he helped to pull it out, or gave its legs a pull with a ratchet on his "calving aid." An hour or two later, he helped the calf to get hold of the mother's teats, if it hadn't worked it out for itself, or milked the cow of its golden colostrum and piped it onto the calf's stomach. It was not clear to me whether the cows worked for my grandfather, or the other way around.

~

One morning he lifted the hand-carved wooden latch, or sneck, on the stable door. The handle was worn smooth from decades of use by rough farm hands and rose softly above the sneck. The rusty door hinges groaned slightly as he pushed the door open about four inches. He peered through the gap. I pushed beneath his chest to see and could feel him tremble with excitement through his coat.

"What's happening?"

"Shush . . . Keep quiet," he said. "She's foaling."

The stable was dimly lit by one half-hearted lightbulb hanging from a twisted gray electric wire decorated with cobwebs from the beams. The once whitewashed walls were brown with years of cattle shit from their legs and bellies rubbing against them. The cobbled floor was six inches deep with white sun-bleached straw. Beneath the dim bulb the bay mare twisted and turned as if in pain. She was staring at her side, which was swollen, some foal limb or other pushing up jagged beneath the taut skin as if she had swallowed a stepladder. Then she lay down in the straw. After a minute or so, a contraction shook her whole shivering body like an earthquake. She extended her head out like a corpse and groaned. Grandad trod carefully toward her rear end, motioning to me to stay where I was. I waited through each contraction, not knowing what was happening. Then she moved a little and Grandad could see a foal's legs jutting out, strangely long, angular, and sharp. He felt past the legs and smiled. The mare shook again and pushed, and the legs seemed to lurch out four or five inches. He waited for the next tremor and then pulled at the legs. This time they came even farther, and a nose was visible, a flash of white seen through the blood and translucent yellowish birth sac. And then the mare stood up, seesawing to her feet, and the foal clattered down to the straw with a thud that made me wince. My grandfather cleaned its mouth of fluids and leaped to the door before the mare tried to kick him.

An hour later, after they had given her some quiet time,

the mare was shaking gently, nervously, as the foal prodded her side with its head, looking for a teat. A week later the foal was running around the field, learning to use its too-big legs and snorting defiantly as Grandad watched it from the field gate. Grandad was still a horseman, forty years after he had bought his first tractor.

~

The previous Christmas holidays I had been sulky and defiant when asked to go out and work. But I was a different boy now. My grandfather didn't ask me to get up when he did at six o'clock: I listened for him dressing and was ready to go when he was. His smile was one of pure pride on seeing me up and ready. Ten minutes later we were mucking out the byre with a shovel. The muck had piled up overnight in a shallow trench behind the cows, landing with dull firm thuds, coiled and dry, and almost smelling of hay. Grandad bent at the middle, pushing his shovel into the steaming piles of muck, then flopped a shovelful up and over into the wheelbarrow. If a cow started to piss, he stood back a few steps to give it a moment. The steaming yellow river flowed to the grate of the drain by the byre door, the whole thing working on gravity. The piss generated clouds of steamy ammonia that made my nose curl.

It was bitterly cold outside, and he said this was the warmest job on the farm. My grandfather worked in the byre in a jumper and waistcoat. The byre was heated until it was snug by the cows and the piles of muck behind them. In the old northern "longhouses," people lived in the same structure as the cattle, he said; this cow-warmth was a precious thing to be made use of. Later, clad in layers of jackets like some down-at-heel Arctic explorer, he went across the frozen fields to feed the sheep, and I bawled and waved my hands to try to keep the hungry sheep from swarming around his legs and knocking him over.

The midwinter work was tough, but my grandfather took pride in it all. He piled up a giant mountainous midden of straw and muck across the yard. He brushed around it each day so that it was neat and tidy. Its steep sides and its brushed-back edges showed that he cared about his work. It gently steamed and was the only place that wasn't being buried in snow, as it generated so much heat that the snow simply melted. He thought that pride in your work, no matter how modest the task, was the mark of a good man, so he mucked the cows out as if he was being judged on it every day. And when the byre was clean, the cattle lying down and full of hay, and the sheep in the field fed, he told me we must use the frozen ground while it lasted to spread some muck on the meadows, without damaging the fields with wheel marks. So, we led some muck out down to the field, with his little tractor and muck-spreader, and it was thrown out by the swirling chains, where it steamed and dozens of rooks scratched among it for grubs.

~

Grandad said we must get a load of turnips for the pregnant ewes. So we took the old Massey Ferguson tractor with the storage box on the back. At the field, we slowly pulled up several hundred turnips by hand and threw them into the box. Pulling them from the frozen ground was hard for Grandad, and almost impossible for me. He puffed and wheezed. But when I said this was "the worst job ever," he shot a sharp look at me and said that pulling them from sodden muddy ground on a cold and rainy day was much worse. I had seen my father in waterproof leggings and jacket working on such days, swollen, frozen red hands running with cold brown mud. After a few hours of this work, it was best to keep out of his way. Cruel work. And yet, for all this trouble, my grandfather and father both loved growing turnips. They were good sheep feed, and the turnip field hummed with life. The rows were a refuge and a larder for wild things when the other fields were cold and bare in midwinter. The frosted leaves sparkled silver in the watery winter sunlight. Hares and partridges and countless other small birds seemed to find food and shelter among them.

A cloud-gray sparrow hawk rose over the dike and scattered the little birds down into the turnip leaves for cover. A big red dog fox skulked out of the far end of the field and away into the woods. As we pulled the turnips, we brought up worms and grubs. Robins and chaffinches snatched them and gulped them down a few feet from our boots. Back at home, in the barn, we threw them out for the ewes to eat. The feast that followed sounded like lots of people eating crunchy apples at the same time.

In spring, these ewes and lambs would go out to help themselves to the remnants of the turnip field. They would rasp at the pinky-orange flesh with their teeth until the turnips were almost carved in two, scraping out every last bit of food and leaving behind a carpet of mud and treacle-black sheep muck.

~

That night the water pipes froze. My grandfather went to and fro from the house, filling buckets of kettle-hot water, carrying them carefully across the icy yard. We poured it down the metal water pipe that rose from the concrete in the byre. After three buckets, and a lot of cussing, the water rattled through and the lumps of ice within shook. The cattle pressed the triggers of their watering troughs, and slurped.

Grandma had a scarf tied around her head like a figure in one of those black-and-white photos of the Siege of Leningrad. She tied old clothes and straw around the pipes and shouted for my grandfather to chop some kindling for the fire. Then she went to make our breakfast. As I kicked the boots from my frozen feet at the back door, I could hear the clicking sound that only well-seasoned logs make when split with an ax. I ran a bath to thaw out, but my grandmother was skeptical; she said people washed too much these days.

~

Grandad's friend George always got the Sunday papers from the newsagents in the next valley, and, after his cows and sheep had been fed, my grandfather would collect them from his house half a mile away. Grandad sat in one of the old armchairs, and George was in the other. I perched on a little wooden chair behind them. The fire glowed orange and the logs crackled and spat. They sat and had a cup of tea and set the world to rights. George was from the same mold as my grandfather. They talked about the wild things they had seen, the price of sheep, who was sleeping with whom in the valley, and who was having money problems. I listened to their talk and could tell that things were changing.

These two old men talked of local people who earned their living from the land on small farms that were now disappearing. Each farmstead in the valley was to them a place where tribes of children were brought up and sent out into the world. Almost everyone we knew could be traced back to a farm—"He's a Weir from Borrowdale," they'd say, resolving everything worth knowing about the person in question. And even as a boy listening to them, I knew that their world of small farms was being eroded with every passing year. All the people in their stories did traditional work, like selling sheep and cattle, building walls, laying hedges, clipping sheep, mending roads, or working in the quarry or the pub. They ignored the incomers to the valley, as if unable to weave these new and different people into their stories. And I would meet many of these (mostly) honest, decent, smart, and kind farming folk with my grandfather when we visited their farms to buy sheep, cattle, or hens. These people lived insular, often deeply private lives focused on their work. Their voices were rarely heard, because they sought no audience. Their identities were constructed from things that couldn't be bought in shops. They wore old clothes and only went shopping occasionally for essentials. They held "shop-bought" things in great contempt. They preferred cash to credit, and would mend anything that broke, piling up old things to use again someday, rather than throwing them away. They had hobbies and interests that cost nothing, turning their necessary tasks, like catching rats or foxes, into sport. Their friendships were built around their work, and the breeds of cattle and sheep they kept. They rarely took holidays or bought new cars. And it wasn't all work—a lot of time was spent on farm-related activities that were communal and more relaxed, or in the simple enjoyment of wild things. My grandfather called this way of life "living quietly."

There was no shame in having very little, Grandad said, quite the opposite. It was better to hold on to their freedoms, even if it meant being poor by modern standards. The constant wanting of shop-bought things he held in disdain.

He thought these people had understood something about freedom that everyone else had missed, that if you didn't need things—shop-bought possessions—then you were free from the need to earn the money to pay for them. You couldn't live from a little fell farm if you wanted foreign holidays and fancy meals out all the time. You had to live within your means.

After a while my grandfather lifted himself from his chair, took the newspaper, and we headed out of the door—pausing to look down the valley bottom to our fields, where our sheep grazed and where the becks met in the floodplain.

~

The grass was silver and crunched as we walked. The beck was edged with ice and sounded as if it was flowing under thin broken glass. The branches on the oak trees were frosted like the velvet on a stag's horns. The sun labored to raise itself above the fell. We were going to meet Dad and John, who had come to lay a hedge. Suddenly, there was a splash in the beck ahead of us. Grandad sped up his walk to a stiff trot. We saw a quick fish-like shimmer in the shallows ahead of us and then it was gone into the dark pools. He told me that the salmon that leave their becks for the Atlantic can taste the rivers they were born in from way out at sea. They follow that taste like an invisible ribbon up the estuaries after years at sea, up the thick slow rivers of the lowlands, on and on, ever upward, over weirs and obstacles, past tree roots and fishermen, on and on, until the taste is so strong they become like torpedoes fueled by adrenaline. Eventually they navigate up the rivers to these valleys surrounded by fells, with their familiar peaty water, and there they wait in the deeper pools for the spate of flood water that will take them farther up the stony shallow becks to where they were born years earlier. And there they will lay their eggs in the gravel, and many of them will die. He told me this as if it was a miracle. He pointed to the fish in the pools, with sea lice on their gills and fins, and white scratches and tears on their bodies, and his face glowed. And as I stood looking, Grandad began mumbling about the "Water Board." He said they were coming to straighten the beck, to drain the valley bottom more effectively, with a series of trenches. They wanted to board the sides with wood to keep it neat. They must be spending someone else's money, because he wouldn't spend his that way. The river would undo their work within a few years, he said.

~

We trekked steadily up from the beck, and through the gate to the meadow, toward my father and John. Dad's Land Rover was already by the hedge and his ax could be heard striking. John was pulling branches away from the hedge and throwing them into piles to burn. Thorns are dormant in winter and can be cut and bent, before coming back to life in the spring. A good thorn dike, Grandad explained, is a solid and useful thing, a simple piece of craftsmanship, requiring no shop-bought things to create or maintain, made entirely from its own materials. He said you could tell whether a farm was thriving simply by seeing if traditional crafts like hedge-laying were still being carried out.

Each thorn trunk was first almost severed with the billhook, then, when pliable and ready to fall, John pulled the branch down to lie on its neighbor. My father protected the delicate sappy hinge, which was thin as a book's cover. His hands were scratched and had dried blood on them the color of the hawthorn berries. He said this thin remnant of trunk would thicken like a wound scabbed over and carry enough sap to the branches to let it go on growing. And as new upright shoots grew from each laid branch, in subsequent years, they would tie the laid branches together until the hedge architecture became tangled and shaggy and thick. The men talked, but I lost interest and went to sit in the car twenty feet away, to warm my feet with the heaters, while listening to Blondie on the radio.

Two or three years later, the branch architecture at its heart was hidden from view, and it was a scraggy green hedge again. Every fifteen or so years hedges like this needed laying, and with each reworking they became an ever more impenetrable, impassable tangle of gnarly, almost horizontal branches. As a hedge aged, it became ever richer in plants, birds, and insects, both in the hedge itself and also on the raised mound, or "kest," on which it stood—a haven for wildflowers.

These hedges were perfect places to hide for us kids, or to climb through in summer. Once through them we were in other kingdoms and could ignore the faraway sound of our mothers calling for us. We would wander off down the hedgerows, on adventures to the railway, or to the old ruined mill where the older kids played—or later, to look at porno mags and to smoke cigarettes. At times it felt like life had always been this way and would continue like this forever.

~

As a child I had had an illustrated book about the Greek myths. I loved Odysseus and Theseus and their heroic journeys. But I suspected we were more like Sisyphus, pushing a giant stone up that hill, only for it to roll back down again and again. At the start of that year of learning I had thought that working on the farm was something to escape from. But I began to realize that, despite some whining and moments of despair, my father and grandfather thought this continuous work was the inevitable price to be paid for a good life on the land. Things must be done, because they always had been done. The secret was to settle in your harness and not fight it. Just get on with it. My grandfather seemed to have found a way to endure it through enjoying the wild things around him, and in taking pride at doing things right. He seemed to be saying to me: learn to see the beauty in mowing thistles, learn to enjoy the skill of the scythe, learn to tell stories or make people laugh so that even the toughest working days won't break you. If he was like Sisyphus, then it was Sisyphus with a smile on his face. He thought, harshly, that modern people were like children, free to play, but bereft of meaning in their lives and disconnected from the things that mattered. He had become stubborn in his old age and suspicious of change, and increasingly sentimental about his ragged old-fashioned fell farm.

My father didn't have that luxury; faced with growing debt, he seemed trapped somewhere between the old farming values and the new economic realities. I felt the tension, but didn't yet fully understand it. That would come later. By the end of that year, though, I had fallen in love with that old farming world. My grandfather had achieved what he had set out to do: I was no longer a boy hiding from the farm; I was a true believer.

~

One fine day, early the following spring, when the grass was a few inches deep and the land dry, the cows and their young calves were let out of their byres and barns. My grandfather untied them and they shook their heads free of their neck ropes, finding their legs again as they walked out of the stalls and byre into the blinding sunlight. Soon they were galloping about the yard and the fields, jumping and bellowing to one another. My grandfather called it "turning out day." It was one of the most joyful days of the farming year. He and the cows were both glad to be largely rid of each other for the summer months. There was no more winter mucking out or twice-daily feeding needed, just checking on them grazing in the fields each day. His daily routine changed massively as those cows skipped away across the fields. The whole farm seemed to sigh with relief. We stood at the field gate and enjoyed them playing like children.

"I reckon they're glad to be getting some sunshine on their backs, and out of that old dark byre . . . Winter is long enough."

~

With the cattle out in the fields for the summer months, the byres fell silent, except for the endless sky-chatter of swallows sitting on the stable door. The barns were transformed from places of noise, warmth, and smells to the coolest, darkest, and quietest places on the farm. My footsteps echoed off the bare stone floors and walls as I ran through them. Beneath the beams the farm cat, Tabby, sat staring upward trying to work out how to kill the swallows. One day, my grandfather saw me looking up at them. He told me about their migration, how their cream chest feathers were still stained red with African dust. As we walked out of the byre, they swooped in and out through the half-open windows to the distant fields where the cows grazed, surrounded by flies.

On our way to the house for dinner, we spotted the birds diving in and out of the log shed. I went over and looked inside for the nest. There it was on a beam. Grandad lifted me up. His arms trembled as he pushed me up high enough to peer in. The mother was chattering angrily on the wires outside. The almost-fledged chicks, with their gaping orange mouths, filled the nest to bursting. They closed their beaks and stared back at me in confusion and wonder. Then Grandad said, with a little urgency, "Can you see them?" and I said "yes" quietly and was lowered back down to the ground.

PROGRESS

Had life gone on like that, then all would have been well. But it was not to be. When all is said and done, our lives are like houses built on foundations of sand. One strong wind and all is gone.

Harakiri (1962)

Only within the moment of time represented by the present century has one species—man—acquired significant power to alter the nature of his world.

Rachel Carson, *Silent Spring* (1962)

Rising clouds of rust-red dust trail me through the night. The wheels of the tractor I am driving churn it up as I race along the dirt road. The steering wheel in my hands judders as I hit potholes. Wire fences and the night stretch out in front of me as far as I can see. A million stars twinkle, like cheap imitation diamonds. I am in Australia. I am twenty years old. I have run away from my home, thinly veiling it behind some words about "backpacking." But whatever I had said I was doing, I am getting as far away from my dad, and our farm, as I can. My grandfather had died three years earlier. While he was alive, it felt as if the old man had cast a spell on us all, making our way of life feel hopeful, decent, and strong, like it would last forever. His unshakable belief in it all made me think we could defy the outside world. I wore that belief like a protective cloak. I was a proud little Spartan. But with his death that spell broke, and our whole world was suddenly exposed and fragile. I could see everything around me breaking and falling apart, and I could do nothing about it. I began to fear that we might be the last generation to work like we did on our hard, northern land.

~

I arrived on this Australian farm—which belonged to a friend of a friend—and was set to work. I was given the job of driving a tractor to some distant land to bale some "Lucerne." I nodded, without actually knowing what Lucerne was. I was not used to working through the night in distant countries on a strange tractor. He explained that they worked through the night because the crop still had its moisture, and that in the baking heat of the day the crop would dry out and be thrashed to dust by the machinery.

~

The tractor headlights illuminate a deep red Martian landscape. Everything is straight. Straight lines. Squares. The farmer told me to head down the dirt track for thirty miles. Then turn right for six blocks, then turn left for two more blocks until I reached the field. It is like navigating on a chessboard. I race past fields of cattle, eyes glowing in my lights. Around them, strange wild eyes are shining out from the bushes. I pass a tree by the side of the road. Creatures half-recognizable are lying around it, large and red. Kangaroos. Startled, they flee through the scrub along the sides of the track. I am so amazed I don't ease off the throttle. They flank the tractor, flying. I could reach out and touch them. It is like a dream of kangaroos leaping around me. Then, before I can process what I am seeing, they are gone and I am suddenly alone in the night, with the vibration of the engine, the stars, and the red dust. And I wonder if anyone at home will believe me about the kangaroos. After an hour of feeling half-lost, I reach the mown field and work through the night, baling by the tractor's halogen lights.

~

The Australian landscape was flat, unlike anything I had ever seen before. It stretched on forever, and then some. A landscape of vast, perfect fields. I was confused by how perfect it was: neat squares carved out of the bush by surveyors with a ruler on a map a century or so ago.

Land here was cheap, the scale vast. There was no history to slow anything down. Or none that was spoken of. It was a blank slate on which these modern farmers were writing the future. No old walls. No old farmsteads. No people. No bones of older things poking out through the new. Just flat fields, perfect for huge machines. Tens of thousands of sheep ranched in fields bigger than our entire farm. Herds of six hundred or seven hundred cows. All the people I met were full of enthusiasm and hope. We can out-compete everyone else in the world, one farmer told me over a beer. He was right. We were beat.

~

A few months later I headed home. I was hopelessly home-sick, and had become crippled by it. I dreamed each night of the fells and the greenness, and our crooked, imperfect fields. And I missed the red-haired girl I'd met in the pub the night before I left.

~

I came back more in love with my home than ever. Our farm had never been so luminous. The hedgerows glowed green as we drove past, the meadows and pastures looked so ragged and pretty, and Dad thought I was talking gib-berish. But I saw, perhaps for the first time, the full beauty of our landscape—its walls, its hedges, its stone-built farm-houses, and its old barns—and I knew that this place was as much a part of me as I was part of it.

And yet, for all this love, I returned somewhat defeated. I had a growing fear in my heart that we would struggle to survive. We couldn't produce food to compete with the farms I had just seen. I sensed that perhaps we were the past, that our age was coming to an end.

~

In the months that followed, I began to see what was hap-pening to us much more clearly. I understood why my old man was struggling to make a living. What good was a farm like ours, battered by the wind and rain for six months of the year? What good were crooked fields and higgledy-piggledy old buildings? What hope was there for farming on this tiny scale? I had seen a new breed of bigger, faster, and more intensive farmers, and now men like them were also emerging from the wreckage of our landscape. I felt embarrassed that my own family hadn't managed to keep up. We were too small, too old-fashioned, too conservative, too poor, and now, probably, too late to find a place in this brave new world.

~

At home, everywhere I looked things seemed outdated. Tractors had replaced horses as the main source of power on the farm a generation earlier. But the tractor implements my grandfather and father used were only slightly bigger versions of what had come before. In our stackyard, the rusting corrugated-tin "implement shed" was full of items of old horse-drawn machinery. The beams in our farmhouse kitchen were still covered in horse brasses. The past had departed, but it had left behind many of its tools, and they were gathering dust in our barns. Ancient tack hung from the beams of our loft, leather crumbling, brittle, covered in white cobwebs: harnesses, yokes, bridles, girths, and horseshoes. There was even a handheld walking seed drill tucked in the beam of our barley loft, with a spreading mechanism that was played by hand like a fiddle. And a pair of horse spurs from the Boer War on the hearth where Grandad had sat to read the newspaper, as if someone had taken them off five minutes earlier. I held those spurs one day and realized that I was part of a traditional working life that was being pulled out of shape. No one on the modern farms hand-pulled turnips or milked old cows for the house twice a day.

~

My father became weighed down with the responsibility of keeping our farm going with all its growing debt. He became rougher and harder to deal with: he seemed to have just one plan—to outwork our problems. Slog it. Get up earlier. Work later. He was irritated by any show of soft-ness in me, as if we couldn't afford to be anything other than tough as hell. He wasn't one for explaining anything in words, but his actions said we had no choice but to copy what everyone else was doing. We needed new machines and new breeds of sheep and cattle; we had to cut costs and corners. We had fallen way behind and now were hopelessly trying to catch up. I was arrogant enough to begin to pity my father for not having all the answers, for not changing things even faster, for not knowing how to win this fight. He carried our growing overdraft on his shoulders like a sack full of rocks. And I became full of a rough kind of pes-simism, cynical and angry. We only had one choice: to em-brace change and modernize. I had come to feel ashamed at how "backward" our farm was. I sensed that history didn't care. It was like a train: it leaves the station, and you can shout "come back" or "you've gone the wrong way" all you like, but it is gone down the tracks and you are left behind.

~

The game playing out on our farm was taking place across the entire British countryside. In many ways, what was happening was "progress." It is easy to forget that farming is literally a matter of life and death; easy to forget how amazing it is to live without ever having to worry where your next meal is coming from—for dinner to always just be there, and better than that, to have a choice of what you eat. And yet hunger was only a generation away for many families in Britain and around the world.

My grandparents had lived through the food shortages and periodic high prices that were common in Britain in the early decades of the twentieth century. This scarcity could be seen in the small stature of the oldest people I knew, who often stood a foot or so shorter than their sons and grandsons. Rationing in wartime was a stark reminder that food could not be taken for granted, when feeding the country required importing 20 million tons of food a year and overseas supplies were vulnerable. These were long years of queuing outside shops for scarcely available goods, cheating and black-market trading for foods as basic as eggs and butter. And so, by 1950, British farmers had been tasked with improving food security and feeding a nation of 50 million people, and were encouraged by government subsidies and guaranteed prices. In the decades that followed they rose to that challenge: producing much more food, much more cheaply. The modern supermarket was the culmination of what people wanted—an eye-popping miracle in historical terms, food in quantities and varieties beyond the wildest dreams of anyone prior to the twentieth century.

In my childhood, my mother's friend Anne, from up the village, was always popping around and telling us how cheaply she had bought a lump of gammon, a bag of frozen chips, or some washing powder. The first big supermarket that opened, fifteen miles away, on the edge of Kendal, was an aircraft-hangar-sized industrial shed, with a huge tarmac car park, and stuffed full of things that were so low-priced it was all people could talk about. Anne would draw at her cigarette and regale us with tales of the things she had bought, or what they had had in the café and how little it had cost. She had stopped baking homemade cakes and teased my mother for being so old-fashioned. She said having a vegetable garden was a waste of time because she could buy everything from the store much more cheaply than you could grow it.

Our vegetable plot was in our farmhouse garden, and it was Dad's job to dig it over and get the potatoes into the ground. He hated gardening. After one of Anne's visits, Dad stabbed the fork into the thin stony soil, and it sang as it hit rocks beneath his feet. The soil in our garden seemed to be the poorest on the farm, as if it were made of handfuls of baked clay rocks. Each spring Dad led in endless barrow-loads of well-rotted straw muck from the calf pens to help it. His spade had a broken shaft and lay by the wall where he had thrown it in disgust when he dug up the last of the potatoes in winter, so he disappeared for twenty minutes to fit a new one. Then, when he returned, he dug a long, straight trench about a foot deep and threw in lots of the muck. I stuck the potatoes on top of the muck and covered them lightly with soil. The potatoes were seeding, because we were a bit late getting them in, and were sprouting with blind white shoots. Something simmered up inside him. He asked Mum how much a bag of potatoes cost in the supermarket. Then he started mumbling calculations about how many hours he spent on growing ours. He declared that growing them was a waste of time. He said Anne was bloody right. Mum countered with a hymn of praise to the fresh garden spud, but he was not having it. Half the buggers were rotten last year with potato blight, he said. That autumn the vegetable garden was sown with grass, and we bought potatoes from town.

~

Over the years, supermarkets began to drive down the prices we received for the things we sold. By the time I got back from Australia, things were becoming desperate. One day we traveled to the local livestock market to sell a load of fattened sheep, with Dad complaining about how little they were worth and that we would be ripped off by the sheep dealers who bought for the supermarkets.

On the way back, we drove past some big lowland farms. Dad was staring miserably at the land beside the road. "Christ, they've given that field some bag." He was stunned by the bulk and color of the crops growing in their fields. The grass was growing insanely fast and was an ungodly dark green—they had been doused in synthetic fertilizer. His comment seemed half in admiration, half in horror—as if he wasn't sure whether this farmer was pushing it all a bit too far.

~

No one in my family was very good at explaining anything we did or offered much by way of clear analysis of what was happening to the farming world around us. So I began to read to try to get answers. I loved the classic farming books, like A. G. Street's *Farmer's Glory* and Henry Williamson's *The Story of a Norfolk Farm*, and I slogged through countless textbooks that were full of useful information but dull as dishwater. I learned that we were a "mixed" and "rotational" farm. Mixed because we grew a number of different crops and kept a few different types of livestock, and rotational because our fields were worked in a sequence that was centuries old.

~

The whole history of farming was really the story of people trying and often failing to overcome natural constraints on production. Chief among these was the fertility of the soil. Farmers learned the hard way through endless experimentation, trial and error, discovering that if we overexploited our soil, ecosystems would collapse, and our ability to live and prosper with it. Fields could not produce the same crops over and over again without becoming exhausted. This was because each crop took certain nutrients from soil, emptying its bank of fertility eventually, and then crop diseases and pests would build up in the tired ground until they became devastating. Nature would punish the farmer for his arrogance. Whole civilizations disappeared because their farming methods degraded their soils.

The solution arrived at from all this struggling and failing was to rotate fields through different crops and uses: some sown with grain, some grazed by livestock, and some left unused and weedy to rest and recover. Different types of crops put different nutrients and organic matter back into the soil through their roots or crop waste after the harvest. After growing wheat, the farmer might sow oats, or, if the soil grew tired of cultivation, rest the fields, leaving them fallow (unused). This sequence helped restore the soil and ensured the fields would feed the crops in the future. My father and grandfather no more knew why this rotation worked than the ancient farmer, but two millennia later they had still abided by the same basic rules. It seemed kind of amazing to me that I could have grown up on a farm, and had eleven years of schooling, and never once had anyone explain to me why these things were done.

As I read through this library of our varied attempts to find farming systems that would sustain us, one of the strangest aspects I discovered was that the field pattern I knew was not timeless—it had once been something quite different. Medieval English peasants had divided their cropping lands into lots of little strips (with the pastures held as commons and each household having a right to graze a certain number of animals). In the cropping fields, each strip was allocated to a different peasant farmer, until he or she had several scattered around the parish, their amount of land reflecting the number of mouths they had to feed. Each peasant grew a range of crops, like oats, barley, rye, and other staples. Fields divided like barcodes would have baffled my grandfather. Why trudge around between strips, when land could have been parceled up in larger, more efficiently sized fields? Why waste time walking between the strips and carrying around plows, hoes, scythes, or sickles? Why leave a wasteful no-man's-land of a foot or two between each strip? But I learned that having these different strips provided useful barriers to slow down the spread of plant diseases and crop-destroying pests, gave homes to pollinators and insects that preyed on the pests, and safeguarded against extreme weather by ensuring each family had their food supply spread in different areas of the parish so that risks like drought or crop disease were mitigated. But the underlying principle was to have multiple crops in rotation, as on the ancient farmer's field—and on my grandfather's land on a different scale. The unbreakable law of the field was sustaining soil health and fertility.

Over the centuries there were major advances in our struggle to fertilize the fields and changes in the ways of parceling out the land, not least in the "enclosure" of those medieval communal strip fields into larger privately farmed fields in fewer hands. In the seventeenth century, British farmers discovered that they could boost the fertility of their soil by planting clover. Clover fixed atmospheric nitrogen (the invisible key to crop fertility) into the soil through its roots and turned the unproductive fallow period into continuing production (previously, this extra nitrogen was only accessible to farmers through lucky lightning strikes). It could be grazed with sheep or cattle (which provided additional harvests of meat, milk, and wool), and this grazing killed arable weeds and trampled muck and organic matter into the turf, keeping the microbiology of the soil working and in good health. (The effect of sheep on tired crop-growing land was known as the "golden hoof" because they made crop land healthy and productive again.) In my childhood, my grandfather still sowed clover with the barley, so that after harvesting the grain the half-green stubble could be grazed by the sheep.

I understood now that my grandfather had fussed over the muck from his cows because it was an important part of the nutrient cycle of the farm—something no sensible person would waste. Everything taken from a field to be eaten by a human, or an animal, took nutrients away from the soil that had to be replaced. I learned that over the past two hundred or three hundred years the ever-increasing demand for food from a growing population meant that many soils were becoming over-cropped and exhausted because plowing (even in a rotational system featuring clover crops) took a toll on the soil and impoverished it over time. No one had told me this. But my grandfather did tell stories about his grandfather using "guano" to grow amazing crops. This dried seabird and bat muck, full of nutrients, had accumulated over centuries in South American caves and at the base of bird-nesting cliffs, and was used as a quick fix from the early years of the nineteenth century—but these ancient natural deposits were soon exhausted. Four generations ago we had been part of the global bat (and seagull) shit economy. But by the beginning of the twentieth century humanity was on a demographic knife-edge. It was believed that without new sources of fertilizer we were heading toward deep and long-lasting famines, as population growth was outstripping our ability to fertilize and grow the crops that were needed. Until, incredibly, a German chemist found the solution.

The dark-green fields we passed on the road carried his signature.

~

No one in my family knew who Fritz Haber was, but without him our lives would have been very different. He had solved the problem of field fertility when in 1909 he worked out how to artificially "fix" atmospheric nitrogen with hydrogen to make it usable for plants. Haber managed to make the impossible possible. He unpicked nature's lock. He had, as he put it, produced "bread from air." Haber's colleague Carl Bosch found a way to apply this process industrially to make things to sell. The resulting ammonium nitrate fertilizer transformed agriculture—and society in general. Some estimates have put the number of humans who could be fed without Haber-Bosch farming techniques at 4 billion (which would mean, if true, that more than 3 billion humans today are only alive thanks to Haber and Bosch).

Haber won the Nobel Prize in 1918 for "improving the standards of agriculture and the well-being of mankind." But his legacy was far from simple or benign. Ammonium nitrate was meant to help feed people—but it was first used as an explosive in the most murderous wars in human history. Haber's other legacy was his contribution to the development of poisonous chlorine gas for use in the trenches of the First World War and the pesticide gas Zyklon-B, used later in Nazi death camps to kill millions of people.

~

From the end of the Second World War, Haber's technological fix spread quickly across the world. Munitions factories in America switched from making explosives from ammonium nitrate to making agricultural fertilizer. American farmers, with access to the new fertilizers, quickly discovered that when their wheat or barley took nutrients from the soil, they simply added (by "top dressing," or scattering on the topsoil) more artificial nutrients and planted the same crop again the next year. It was no longer necessary to use "circular" nutrient systems, with a diversity of rotating crops and livestock, in order to maintain a healthy soil. These farmers were freed from having to create their own fertility within the farm and offered a miracle shop-bought solution. Haber's nitrogen didn't change all farming overnight—it would play out over decades—but this was the first profound break with the past from which so much else would flow.

In the postwar years, salesmen for the chemical and machinery corporations brought artificial fertilizer to rural communities around the world, including ours. They would come to our house for tea and cake, handing out glossy brochures with photos of amazing crops, and blather on about all the things we needed to buy to keep up with our neighbors.

~

A couple of weeks after we sold our sheep at the auction, I was sent to "lead silage" for my dad's friend who owned the dark-green fields. I was put on a giant green John Deere tractor. The red trailer towed behind held about ten tons of grass. REM and the Smashing Pumpkins blasted and whined from the radio. Half a dozen of us had the task of leading grass from the "forager," which hoovered up the juicy green grass and chopped it in the field, back to the silage pit, or "clamp," in the farmyard. At the pit we had to tip the trailer, spewing the wedge of grass onto the concrete, and then race back to the field for another load. Weeks earlier, in the spring, Dad's friend had sprayed, "top-dressed," millions of tiny polystyrene-like balls of synthetic nitrogen onto the field with a tractor-driven "fertilizer drill" and powered this grass into life. The grass was dense and deep and utterly uniform—all one species, high-performance ryegrass sewn a year ago—and it thrived on the new shop-bought fertilizer. The field grew so quickly that not only were we harvesting the grass three weeks earlier than was ever possible in the past, but it would regrow for as many as three or four cuts that season.

Silage is just pickled, moist green grass, but when it first arrived it was a kind of farming miracle. Unlike hay, it didn't need sunshine to dry it and could be made in one day, even if it was raining. It made more nutritious food for cows, which then produced more milk or meat. The value of the improved nutrition of silage, relative to the hay that might otherwise be made on a farm like this, was valued in the tens of thousands of pounds.

That afternoon, we packed hundreds of tons of grass

into the concrete and steel-sided silage clamp every hour. Our friend Rusty pushed it in with his tractor and a giant buck-rake, load after load, and rolled the air out of it. Just before it went dark, we sealed the pile of grass with a huge plastic sheet to make it ferment, and threw hundreds of used tires on top to hold the sheet in place. We had a can of beer each, all of us proud to have got so much grass safely harvested in a day. When someone asked me when we would make ours, I felt embarrassed when I told them we might be another month. I didn't mention that we still made lots of hay the old-fashioned way—it suddenly seemed prehistoric.

~

Nothing about that day of making silage was entirely new. The improved grass, the synthetic fertilizers, the tractors, and the practice of making silage all had roots three or four decades back. But the scale and pace were different—it was all much more intense. I was experiencing the latter stages of a race almost run.

~

I came home from Australia full of bright ideas about how we could modernize our farm. But Dad just shrugged and walked away, perhaps too proud to say out loud that we hadn't the money to start buying fancy new tractors and trailers, and perhaps not entirely convinced he wanted to anyway. He seemed a bit stranded between the old ways and the new. He made do with an aging tractor or two, and some rusting secondhand machinery, and old hay barns. We did make some silage, but now he thought we should continue to make meadow hay as we always had.

So we waited for a weather forecast of four or five fine days in July, and then started the cycle of mowing, drying, turning, baling, and leading into the barns. It was a nightmare that dragged on for weeks. Our first attempt at hay was ruined by rain, so after that we took it a field or two at a time. We fiddled on slowly, trying to get some made and into the barns safely, to minimize the risk of a winter forage shortage. It was weeks of sweaty, dusty manual work in the fields and stacking bales in the barn. That meant recruiting my elderly uncle to drive the bales to the bottom of the elevator with a mechanical clamp, while my mother lifted them onto the petrol-driven elevator, and my father and I were up in the eaves, throwing them to each other and laying them down in layers, in the stack we called a "mew," crossed at the joints, where they would hold each other in place until winter. Honestly, I didn't mind the work, but I was frustrated by how slow we were compared to farms with more staff or machinery to handle the bales. I mocked my dad, saying we were like *Dad's Army.*

~

We might not be able to afford the latest machines, but I knew from the leaflets left by the traveling salesmen that if you really wanted to crack this modern farming game you needed to use the latest pesticides.

The spear thistles were waist-high, their purple flowers beginning to open. The bottom acre of our Cow Pasture had been taken over by them. It had become so rank that even the cows barely bothered to go there. There weren't enough people left on our farm to scythe these weeds. They were getting out of control, and someone had to do something. I thought we were hopelessly outdated, and I was fed up that the changes my dad made still weren't big enough or fast enough to catch up with the large modern farms. So I bought the latest thistle pesticide from the agricultural merchants we used. It was cheap. It came in a squat brown plastic bottle and I diluted it in water. I also bought a mini-sprayer that was mounted on my back with straps, called a "knapsack," and spent every night for a week zapping every thistle and nettle on the farm. I swung the lance backward and forward until I was surrounded by a white cloud of spray, and the back of my mouth felt dry and tasted bitter. I mixed the stuff twice as strong as the instructions said, because we all knew that the "boffins" always played on the safe side. Satisfyingly, the thistles and nettles were wilting by the time I passed home, their silver underleaves turned over.

A couple of days later they were black and shriveled. There was hardly a living thistle or nettle in the fields. Not only did it kill every thistle, but it also stopped them from seeding as well. I went back and sprayed any survivors until the field was clean. Within weeks whole areas of our farm that were once plagued with thistles could now be cleared and we were able to grow grass without weeds. Work that had taken three of us several days every summer no longer needed doing. Our scythe became another museum exhibit hung in the roof of our barn, growing rusty with each passing year, triangular cobwebs stretched between blade and shaft. Our fields started to look much tidier, and more like the neat modern farms we knew. I sprayed the nettles in the stackyard, where the old machines were left to rust, and the edges of the fields, where clumps of nettles and other weeds grew. This was modern farming. The spray was bloody miraculous stuff. Even my dad was won over. We were heading for the future.

~

I knew we were late to the party. Synthetic pesticides had been saving crops from weeds and diseases since the Second World War. In 1939 Swiss chemist Paul Hermann Müller found that dichlorodiphenyltrichloroethane (DDT), a chemical developed in the 1870s, could be used to kill insects. DDT was, like ammonium nitrate, a Nobel Prize–winning "miracle." It was used with great success to eliminate typhus almost completely from large parts of Europe during the Second World War by killing off huge numbers of mosquitoes; it was also used to eradicate malaria in America in a few short years after that. DDT was marketed to farmers as an insecticide. It could be used to kill almost any insects, fungi, bacteria, or vermin that destroyed or spoiled crops, as well as insects that carried pathogens that could make farm animals sick.

I remembered my father looking utterly defeated by mildew blotches on the leaves of our barley and how sickly and wasted it looked. All our crops were vulnerable like this. The infamous Irish potato famine, which was triggered by potato blight ravaging a society over-reliant on a single crop, wasn't a one-off: it was a particularly disastrous (1 million people died, and a million more emigrated to escape it) version of something everyone lived in fear of prior to the advent of pesticides: failed harvests and rotting crops. Being able to grow huge quantities of nutritious affordable food to "feed the world" without plagues of insects and crop diseases was revolutionary. Pesticides and artificial fertilizers offered us amazing new tools to do our work more efficiently than we ever could before. Vast amounts of food were lost to pests, diseases, and vermin, or through bacteria or mold, both in the field while being grown and while being transported, stored, or sold in the shops. The chemists were freeing agriculture and the human food chain from its eternal natural constraints.

~

My father spent the following summer in a blue T-shirt, on a tractor, with a giant milky cloud of pesticide rising behind him. Our old tractor trundled up the barley field, half submerged in the yellow sea. On the back was a crop sprayer with wide booms, arms that reached out for about twenty feet on either side of the tractor. My old man was going to grow a "clean" crop of barley like everyone else. The age of weeds was over.

As I watched him, I recalled how my grandfather had oozed contempt for tractors. He used them, and the machinery that came with them, but he didn't like what they did to us, because the moment we stepped up onto them, we raised ourselves from the earth, no longer touching it, smelling it, feeling it. That sensory contact was the essence of knowing the land. Now we were spending more and more time on tractors, encased in glass, steel, and plastic, and distracted with air con and the radio. He thought machine work was of a lower order of importance than working with animals or with your hands. Any fool could drive a tractor around and around a field. But tractors vastly increased the scope of what we could do in a day, and we had little choice. The number of people left working in the fields was declining, and the amount of time the surviving farmers spent in the fields was reducing too.

In the 1970s Grandad had had a 45-horsepower Massey Ferguson. Now our tractor had 100-horsepower, and some of our friends had 200-horsepower tractors. On the bigger farms the giant tractors could now rip out a tree by its roots as if it were a toy. We watched in awe as the new machines on those farms were used to achieve incredible efficiencies of scale, reengineering landscapes entirely. There was little sentimentality. Fields could not be museums. Small, narrow, or crooked fields created in the age of the horse were no good now. Trees, rocks, thorn bushes, walls, and boggy areas got in the way of tidy machine-work. Fields were quickly made larger, flatter, better drained, and less weedy. Obstacles like woodland, hedgerows, ponds, bogs, or rivers were cleared, drained, straightened, or filled in. There could be no barriers to machine-work. An efficient landscape was required to make food cheap. A new combine harvester might cost half a million pounds, and so it needed to cover the greatest possible number of acres to earn back its cost rather than messing around in little fields.

~

On our dining-room wall hung aerial photos of our farm, taken from a light aircraft by enterprising photographers. They sold my family one of these farm pictures every few years. On them you could see how the fields and farmstead were changing size—not overnight, but steadily, a little bit at a time. By examining those photographs over time, you could see how the field boundaries were disappearing rapidly—walls, hedges, and fences vanishing. On our rare excursions to southern England for a sheep sale or an agricultural show, we would stare out of the car windows, amazed at whole landscapes already simplified to grow one or two crops intensively from horizon to horizon. What worked at a farm scale could also now be made to work on a much larger scale, with whole regions specializing in doing one or two things. A traditional mixed farm had limits to its scope: it needed a farmer and people to move cattle or sheep around, lay hedges, build walls, and do other craft work. The new farming needed none of that.

~

My uncle and aunt went on holidays to North America to see the latest high-performance livestock and machinery at first hand. They had a better lowland farm than we did and were way ahead of us in embracing change. They came back from America as evangelists for change. They brought back baseball caps and Hershey bars. We listened as they spoke of even more advanced tractors, machines, and livestock, and soon these things were appearing on local farms. To the eyes and ears of people who had spent years digging drainage ditches by hand with a spade, or who had followed behind horse-drawn plows with frozen hands and feet, and scythed meadows until their hands blistered and their shirts were wet with sweat, these felt like giant steps forward.

~

My mother was sobbing in the kitchen, tears flowing down her face in front of everyone. None of us had ever seen this before and it filled the room with confusion, and another sensation I hadn't experienced before: shame. The whole family had been to a farmer friend's funeral, and afterward there was a family row about something that seemed to be about everything but the stated cause. My father had apparently let the family down by not having the right kind of funeral coat. The men are meant to look respectful—and all the same—in thick woolen coats, dark gray or black, inch-thick to turn the grief, while they stand at the back of the church, or in the churchyard, singing old hymns about shepherds. I had no idea there was a right kind of coat, but it started a feud in our family. One of Dad's sisters said she felt ashamed he wasn't dressed properly. She said he looked scruffy and cheap. He wore a thinner coat, because he didn't have a woolen funeral coat. He had broken the rules of respectability and given people a glimpse of our reality that they shouldn't have had.

My parents and grandmother came back to the farmhouse. Dad sat silently in his chair by the fire, looking wounded and angry. I gripped my cup of tea. I wanted Mum to stop crying but didn't know how to help. My grandmother was fussing about something in the cupboard that didn't matter at all, desperate not to step between her daughter's words and her daughter-in-law's tears. Mum was slumped at the kitchen table. She found this farming life full of strange rules and confusing customs. She wasn't used to these huge communal funerals and what it all meant, and the symbolism of it. She didn't show emotion very often, but this was all too much. She said they "cannot win." It was just a fucking coat. And I was aware suddenly that this was all about money. Dad didn't have a proper coat because my mother was trying to save money. The money my parents borrowed to buy the land on my grandfather's old farm years earlier was spiraling into a huge debt because interest rates had gone insane. Mum said they were damned if they spent money, and damned if they didn't. She knew she had failed, that they were meant to present a front to mask it all. It would be some years before my father forgave his sister. For years afterward he would glare at whatever clothes we wore to funerals to make sure we were dressed properly.

~

We had regular visits from the bank manager. Mugs of milky tea were presented, with three plates of cakes and biscuits. Box files of bank statements and bills would appear on the table. I was sent out to work—out of the way—while they talked. But the silent glances exchanged between my parents at the next mealtime were enough to tell me that things were really bad. Grandad had been willfully blind to the world we lived in, and defiantly held on that way until he died. Dad couldn't afford to be like that. He said we had to do something before we went bankrupt.

~

I found books by economists who seemed to understand the rules of this tough new world. I was hungry for what they knew. I admired their realism, even if I also knew it meant a cruel end for the world I had loved. Joseph Schumpeter had seen it all coming, way back in 1942. The death of small farms was not only inevitable, I read, but a good thing for society. He described this as an inescapable capitalist process, what he called "the perennial gale of creative destruction." No one likes being on the wrong side of history, said the economists, but get over it. Small farmers were like the coal miners, yesterday's people. Go back to school, retrain, move on. President Nixon's secretary of agriculture, Earl Butz, had famously told audiences of farmers, again and again, to "get big or get out," and to plant commodity crops such as corn from "fencerow to fencerow." He said the old ways were hopeless and needed to be eradicated: "Before we go back to an organic agriculture in this country, somebody must decide which 50 million Americans we are going to let starve or go hungry."

While we worked, I would regale my father with all this economic thinking, and he would sigh heavily and say nothing. I knew we needed to be more "productive" and more "efficient." We needed to be more selfish and stop wasting so much time on things that didn't make any money—Dad would volunteer to tidy the village green, paint the village hall, or help a neighbor with their sheep shearing or with their haymaking. He loved breaking in young horses and could spend hours doing it. He loved going to the auction mart to follow the trade and buy and sell a few sheep or cattle.

~

The new technologies and ways of using them unraveled our farm like someone pulling at a loose thread on an old jumper. First the horses disappeared. The pigs vanished next. Then the small flocks of turkeys and hens went. As pieces of the farm were taken away there were all kinds of knock-on effects. When the horses were sold, our need for fields of oats went too. When the dairy cattle went, the milk stands grew nettles and the butter churns, paddles, and molds were stashed on a dusty shelf in the pantry. We no longer grew fields of turnips or barley. Instead, we bought cheap sheep feed from a local mill made from imported American lupins, maize, or palm kernels. All the fields on Dad's rented farm were soon a single shade of green.

Across the country a great simplification had already taken place, and it was speeding up, as we all tried to keep up with the new intensive farming by copying it. I was witnessing the climactic stages of a great stripping away. Farms shed layers of rotation, specialized in certain crops and animals, applied artificial fertilizers and pesticides, bought new machines, and used whatever other ideas and inputs were available to increase yields and keep up. It was a kind of arms race, with the large modern farms trying to swallow up the little old-fashioned ones like ours.

~

The economics of farming made it virtually impossible to opt out. There wasn't an option where farmers got to press pause and stick at some moment in time just because they wanted to. They'd simply go bankrupt or fall into a spiral of debt as we were doing. Prices for farm products were now global and were being driven down by the vast quantities of commodities being produced by the super-efficient new farming that dominated in North America, which was spreading rapidly around the world. Europe tried, often ineffectually, to stop or slow these processes with greater regulation and protectionism—they had an alternative version of farming, with different rules on things like animal welfare and antibiotic use—but over time many of the same problems emerged. The new farming undermined the old farming systems by undercutting them. Prices no longer reflected local or seasonal farming realities. In real terms, our sheep were now half or less of the price they had been a few decades earlier. Giant ships full of frozen sheep meat now arrived in British ports from New Zealand whenever the price threatened to rise, and drove the price down again.

~

A little Welshman with a moustache who worked for the government came to see us once a year. He encouraged us to do more and more to claim government grants to "improve" our farm by making the fields bigger, draining the bogs, and generally making it all more "productive."

~

I would sometimes be sent to work on some of the biggest farms in our area and saw what the new farming looked like up close: giant weedless fields, huge machines, and enormous buildings. The farms I had known growing up had been slightly ramshackle because they slowly added new buildings to the old. The effect was a bit like a hermit crab, shedding a series of outgrown shells. In the 1950s farm buildings in our area had broken out of the old laws of 22-foot spans (the span possible with local trees for beams). They grew with each decade, and now a few miles away the farmers were building monster sheds and giant industrial complexes made of huge steel beams and concrete panels, with spans of 80 feet plus, and 250 feet long or more. Soon there were farms that looked and worked as efficiently as factories. A pig farm we knew, down in the Eden Valley, bulldozed away the lovely old sandstone barns and stables with their beautiful arches, leaving heaps of rubble and just a stone pad to build the new sheds on. My father thought this was an act of mindless vandalism, symptomatic of the new farming in which nothing old was respected. But to the new way of thinking these old buildings were just in the way.

Ten years earlier I had played with the ginger-haired boy who lived in the cottage on that farm. It had five thousand pigs. We roamed through the sheds past the farrowing crates, with sows grunting and jangling the chains. We lifted the dusty lids of the pigpens and gawped at the squealing piglets under the orange heat lamps. The ammonia from the pig slurry made our eyes weep. Mice were everywhere. We were told to "piss off" by a man with a cigarette in his mouth, who was the "pigman." By the 1990s that farm had close to 120,000 pigs and sold more than five thousand to a supermarket every week. The farmer had a fleet of trucks— for hauling pig feed in and transporting pigs out to abattoirs. Looking back, I can see he didn't really have much choice. The new farming effectively devalued a pig to a fraction of its historical value and shrank the profit margin on any individual pig to such a tiny amount that only vast industrial corporations could afford to produce them—no modest-sized farm could compete and survive. My father had once kept fourteen sows that produced about one hundred fattened pigs a year for sale, but on our farm, as on thousands of others, small-scale pig-farming was given up, replaced by a handful of industrial-scale pig units elsewhere. The same happened with chickens, as they too were easy to industrialize and could be fed on the cheapest grain. And through it all we just adapted and bent to the will of the world.

~

I am not sure exactly when things that originally seemed like a good idea began to feel like a step too far. I can't pretend I had any great wisdom. All I can remember are brief moments when I first saw my dad's doubts grow, or when my own faith in the future began to falter.

For years I had been aware of the ugliness and the strain. We had said goodbye to skilled farm workers, been squashed down by farmgate prices, and scaled up our livestock beyond what I would once have thought possible. We knew farmers who couldn't hold it all together and were drowning in work and debt and chaos. It was becoming impossible to ignore the farms falling into squalor.

The economics books I read were all about how things changed for the better; they didn't say much about the losers, the misery, and people hanging on for years, sometimes decades, because they knew nothing else. Our community was fracturing and breaking.

~

The only trace of the cow that was left was a few hoof and leg prints on the crust of the slurry pit, and then a dirty churned area where she had fallen through.

We had trekked across the fields on the trail of some of our cattle that had escaped from a field we rented a few miles from home. They were two-year-old heifers, skittish and fat, with curly red coats. They had pushed through an old hedge and wandered across some rough ground, through a second hedge, across a big field, and then had found their way into one of the big new dairy farms. We got most of them back safe, mended the hedge and then came back to see what had happened to the lost one. Dad soon worked it out. She had run into the farm's slurry pit, and after trying to gallop across the deceptively firm surface had fallen into the deeper part and sunk down into the darkness. Dad was frustrated and angry, because the lagoon wasn't fenced properly and was clearly dangerous. He seemed to want to climb in himself to be sure his cow was in there—but thankfully seemed to think better of it.

The farmer turned up and was defensive, said our cow wasn't in there. Dad told him to stop being bloody ignorant; she would come out when it was emptied. He told the man that if he'd done any fencing our cow wouldn't have got this far. He should get a safety barrier around his slurry pit before some kid fell in it and drowned. On the way home he ranted about the mess on some of the farms now, and how dangerous they were. "Those places are bloody lethal," he said. It was like we were all being sucked into a whirlpool, and although we might try to swim faster and faster, we were all slowly becoming exhausted and being drawn into the darkness.

~

We were standing on the roof of a small shed, ripping the timber boards off the barn with a claw hammer and a crowbar. It was a mild cloudy day and the boards were dripping with condensation. Steam escaped from the hole we'd opened. We were desperately trying to let more fresh air in after an outbreak of pneumonia, which thrived and spread in warm damp places, like our overcrowded cattle sheds. Each autumn my father bought year-old "store" cattle in our local markets. In summer they would graze out on the fields, and in winter they would be housed in barns like this one and bedded with straw. They were new "Continental" breeds that grew fast and produced meat quickly. Red and black Limousins that were as wild as hell. White Charolais cattle that stood about a foot taller than some of the old native breeds and piled the weight on if you fed them well. Black-and-white Belgian Blues with extreme double-muscled backsides. Dad grew them, wintered them, fattened them, and sold them to butchers. After a while we converted the disused hay barns into a loose shed for cattle and fitted gates so that they could reach through for silage and bought-in feed. And every year we had more cattle and less space, because the margin was shrinking with every passing year, so we had to cram more and more cattle in to earn the same return. But this brought its own problems, because too many cattle in a barn would get dirty as they trampled over their own bedding; and in those badly ventilated converted buildings, where the air could quickly become stale, they would get diseases that spread in cramped conditions. Sometimes we would have a disaster and lose two or three bullocks and the profit on the whole lot would vanish. As good as these cattle were—and they were beautiful and amazingly productive—they weren't as tough as the native breed cattle we had kept in the past.

We also kept two or three times as many sheep as my grandfather had. My mother and I lambed them when my dad went to work on what had been my grandfather's farm. We shifted to more modern, "improved" breeds that grew quicker, with better and more valuable carcasses, but they were lousy mothers, ate more shop-bought feed, and often died for no apparent reason. At lambing time, there were not enough hours in the day. The barn was divided up into pens made of straw bales and pallets held together with bale string. Mum would rush around them, making sure any weak lambs were fed, filling water buckets, giving them hay, and bedding them to try to keep them clean, but there were always too many in the barn, and not enough of us. Then, when the rain stopped, we would lead the strongest ewes and lambs out to the fields in a trailer and make sure the young ewes mothered their young properly. I would help Mum catch any ewes giving birth, grabbing them, because she had a pin in her ankle from a bad break years before and after a hard day she would be lame. My father would make her angry by coming home tired and hungry and critical of our work with the sheep, saying the weather was worse in the fells. He would sit at the hurriedly laid supper table and say, "I haven't got a fork," and my mother would look as if she might stick one deep in his chest.

~

It was becoming clear to me that the way we thought about and interacted with our animals was changing. Our farm animals had never been pets, and we were rarely sentimental about them, but there was a lot of care involved and a kind of intimacy that vanished as things got bigger and more industrial. The farm animals' characters were well known to us: they all had backstories. To earn the title of "stockman" or "stockwoman" we were expected to have an encyclopedic knowledge of all our cows and ewes. But on the growing modern farms something had changed: it wasn't necessarily that animals were ill-treated—I didn't see much of that—but more that they had just become units of production.

For most of history, animal products like meat, milk, and eggs had been expensive because they couldn't be produced with industrial efficiencies of scale. The logistics of feeding thousands of pigs or hens, or cattle, would have defeated a farmer prior to the modern age. Most of what animals ate was grown on the farms, and for most of the year was harvested by the animals' mouths. This limited the scale of the livestock farming to the animals that the farmers were able to keep through winter, either on the hardy crops they could grow, like turnips, or on the harvested crops they could store in the barns, like barley, oats, and hay. But now there was no limit, as tons of cheap feed were only a phone call away. Farmers could readily scale up using massive buildings where environmental conditions could be tightly controlled—and the conversion of feed into bodyweight, milk, or eggs could be made much more efficient. No self-respecting modern farmer would keep an old or ailing cow the way my father had done with Old Blackie. The pigs, chickens, or cows in those giant sheds didn't exist as individuals and had become more like a crop, a mass-produced entity generating a "yield." Perhaps it didn't matter to most people, but I found it unsettling and alien.

~

On the modernizing dairy farms that we knew, the cows no longer went out in the fields at all. Once a farm had more than around two hundred cows in the herd, the logistics of them going outside became problematic. They took forever, holding up commuting traffic on the lanes and roads, plowing up the field on wet days, and trampling and spoiling the grass through the sheer weight of their numbers. Mowing the grass and leading it to them in a trailer to the buildings was more "efficient." The cows didn't "waste" calories on walking. Logistically it was sensible, almost inevitable, but my father didn't like it. It was awkward to speak about it openly, because our friends did it. Supermarkets were advertising low-priced milk as part of a price war at that time. Dairy farmers couldn't afford to stand still as the real price fell—the price of milk was lower than that of bottled water. These dairy farmers had to get bigger and intensify. They just about convinced themselves that it was OK. The cows looked well in the barn, they said, and they did their best. My father was no animal-rights activist—his own cows stood in stalls or barns in winter, sometimes they looked a bit dirty, and he accepted the realities of farming cattle—but denying cows the chance to spend their summers in the fields and that moment of freedom in spring when they ran off down the fields and played was a break with his ideas about what was right for a cow. But the new farming had created its own morality and ethics, and the people caught up in it had to change theirs, or get out, and things that initially shocked them soon became the new normal.

~

The history books I read made it very clear that this wasn't normal at all. Prior to the twentieth century, keeping farm animals in large numbers in one pen, barn, or field for anything more than a short period of time was courting disaster. Housed animals failed to thrive or became ill, like our bullocks had done that winter, because the dirty conditions led to outbreaks of disease and parasites, and because they simply didn't get the vitamins and minerals they needed. So now farmers trying to pursue intensive methods of animal production were prone to suffer catastrophic losses.

In the wild a lot of parasites live on animals like cattle, sheep, pigs, and chickens, but their ability to devastate flocks or herds is minimized by the fact that the animals are spread out over the countryside and interspersed with other species (keeping the jump between host animals minimal). There is limited transmission of disease through saliva, urine, or muck. Free-grazing wild animals tend to move themselves (or are forced to move by predators) away from land that is covered in their muck to find fresh pastures, so they have less exposure to parasites like intestinal worms. Traditional pastoral systems tended to mimic what worked in the wild: grazing cattle or sheep were healthiest when they were either herded around a range of habitats by a shepherd or cowherd, or left to their own devices to roam across whole landscapes. They helped themselves to an extensive range of plants that gave them both the diet they needed and also the minerals and vitamins, and in some cases the medication too.

The new intensive farming placed animals in surroundings that made them dirty, stressed, and diseased, and then used medicines, particularly antibiotics, wormers, hormones, and vaccines, to cure those problems. Animals could be kept healthy with antibiotics in crowded, industrially scalable conditions that would once have made them ill. Antibiotics were, of course, used to cure individual sick animals; but more worryingly they began to be administered to large groups of animals to prevent illness and, surprisingly, to promote growth. In 1950, scientists in New York had discovered that by adding tiny traces of antibiotics to animal feed they could increase the growth rates of animals. Thereafter antibiotics were routinely also used in animal feed, particularly in the most intensive American systems, for cattle, chickens, and pigs, to improve feed conversion.

Behind the antibiotics and vaccines came a whole host of other medical products, such as anthelmintic wormers used as drenches, squeezed down the throat to expel internal parasites, pesticides to kill external parasites like lice, hormones to make animals grow quicker, organophosphate dips for sheep to kill wool and skin parasites. With these tools, farmers could now concentrate animals in confined areas on a scale that had never before been possible. Farms were becoming machines. It was farming by numbers, as designed by accountants. Critics called it "factory farming."

~

Dad would occasionally be hired by a big dairy farm to help with their grain harvest. He'd drive a giant tractor pulling a giant trailer, leading grain from the field to the barn, and he saw how the cows were being farmed. He hated the sloppy, couldn't-care-less work ethic and frequently came home grumbling. He said the cows on that farm might give ten gallons of milk a day, but they were living on a knife-edge and giving far too much of themselves to be robust. Unable to cope with bad weather and disease, they had to be mollycoddled like thoroughbred racehorses. About one in ten was lame at any moment in time, with sores on their knees and hocks. They were lean as old crows, and hobbled about with huge swollen udders, prone to mastitis. The proud old cowman who had once doted on these cows had left, after years of complaining that the never-ending growth of the herd meant things were being done badly.

One day a cow gave birth; it didn't seem to be anyone's job to keep an eye on it, and hours later the calf was needlessly dead. Dad was angry and confused. He didn't understand how this could happen. The boss's elderly father arrived later and cursed at the dead calf lying on the concrete. It was obvious that no one was looking after these cows the way he had once proudly done. He said the men were "useless fuckers," but he also seemed demoralized by this new system that had outgrown the former ways of caring. He was now too old to do anything to make it right. The men just shrugged and skulked off to another job. In my father's eyes the work, the land, the cows, and the people were all being devalued. Traditional farming people were being broken in spirit. A farmer's pride in seeing and judging things carefully was dying out. When my grandfather had stood looking over a gate, he was figuring these things out by close and thoughtful observation. Managing animals the traditional way required specialist knowledge and judgment, and skilled people to care for them and understand their needs. These skills didn't scale up easily for mass production; such farming couldn't be uniform or predictable. Animals came in different sizes and shapes, and they matured at different times. Farmers slaughtered, preserved, and cooked animal meat when it was available and ready, not every week. It was all a long way from the factory ideal of identical commodities all being ready for the shops at the same time. So, incredibly, farm animals had been made more uniform.

Some of the new super-intensive farmers were our friends. And when I talked to them in pubs or at agricultural shows it was clear they operated with an entirely new way of thinking. They were like a different species of farmer entirely. They applied science, technology, and engineering to solve farming problems, and made it all work with industrial efficiency. They were the economists' ideas incarnate.

Scientists working on animal genetics were able to identify and strip away "useless" genetic traits, and that included some fairly basic attributes and instincts that animals had always needed in a semi-natural setting. The focus shifted to developing productive traits like speed of growth and body bulk, improved milk yield and feed efficiency. The bits of the animal's body needed for movement, or foraging, or even natural reproduction could be shrunk with each new generation, and the parts that were valuable for human consumption could be grown. These changes to the physiques of farm animals didn't come in one fell swoop, with a single magic ingredient—they came by the application of a whole suite of scientific disciplines and tools used in combination to find "marginal gains," not unlike the way an elite sports team would be put together. But the transformation in how animals looked was remarkable and unsettling.

The greatest productivity gains were in pigs and chickens, which could be housed in great numbers, able to reproduce at very rapid rates, and selectively bred to convert cheap corn or wheat efficiently into meat. I read that since the 1950s the time it took to get a chicken from the egg hatching to slaughter in the most intensive systems had reduced from sixty-three days to thirty-eight days. The feed needed per chicken halved. At the same time these new pigs and chickens were being kept alive with antibiotics, fed heaps of protein, and housed at constant temperatures. Farm animals have always been exploited (it isn't considered a very nice word, but it is basically true—all cellular life depends on using other organisms), but in these intensively raised animals it had been taken to the extreme.

Large corporations engineered these changes and then "owned" the genetics of the improved pigs or chickens, as well as the supply and processing chain. Chicken farming had been taken over almost entirely by big business, and small farmers disappeared or were reduced to farming pigs and chickens "on contract" for the large companies. By the time I was twenty there were barely any chickens or pigs on any local farms, except on one or two giant industrial pig units a few miles away.

Many of our friends and family were dairy farmers, so we saw that world change at first hand. In my father's childhood most dairy cows in our landscape were Shorthorns: red-and-white, or roan, thickset cows that were fairly tough in all outdoor conditions and were "dual-purpose" in that they produced beef and milk, at sensible, but not extremely high levels. Historic records give us a very accurate baseline for the productivity of a cow under the traditional farming systems. The Shorthorn yearbooks for the breed tell us that in 1954 through 1955, the largest herd in our area (with recorded milk yields) had thirty-three cows. The Shorthorn Cattle Society awarded medals for the highest-yielding animals, those that were producing three or four gallons of milk per day. Dairy Shorthorns were first replaced by black-and-white Friesian cows between the 1960s and 1980s, and then by North American Holstein cattle from the 1990s. These heavily engineered cattle produced more than twice as much as the cows my father milked in my childhood: nine or ten gallons of milk per day. It is worth pausing for a moment to process that. It took ten thousand years of domestication and gradual selective breeding to create a cow that gave four or five gallons of milk per day, but in my lifetime that amount has doubled. Few people outside farming have registered how incredible this change is. The new highest-performing cows often last only two or three lactations (milking cycles after each calf) before they are worn out—suffering from lameness, mastitis, or simple exhaustion from being too specialized in pursuit of producing too much milk. Dad held Holstein cows in contempt; he said they could hardly be out in a shower of rain without catching a cold.

Remarkably, this process of intensification is still speeding up, not slowing down. The largest British herds now have more than one thousand cows, and some herds globally have tens of thousands. More than 50 percent of British milk is now produced from cows that live permanently indoors. The cows are changing so fast in the elite herds that only the first daughters of each generation are kept for the herd, because by the time they calve for the second time, a year later, the genetics of the next generation of cows (a year younger) are superior to theirs.

~

As the animals disappeared indoors, the large gangs of farm workers that had once played games of football on our flattest field vanished. The last of our live-in farm workers, Stuart, had left his little bedroom in my grandfather's house to live in the local town in 1978. He had been like an extension of the family and was nursed by my grandmother when he had cancer. He had been even more of a teacher to my father about practical farming matters than my grandfather had. Such men (and women) were everywhere thirty years ago. They knew the fields intimately, sometimes better than the farmer. But fewer people worked in the fields with every passing year. Now, most people would never set

foot in the fields that fed them—which was either libera-
tion from a kind of mindless drudgery, or a loss of contact
with the vital processes that sustain us, depending on your
point of view.

~

Our lives were becoming more fragmented, more private,
closed off behind doors. The local dances that my father's
generation had known no longer really took place. Drink-
driving laws in the 1980s had mostly stopped the journeys
that made many of these social events possible. The pub
in our village had closed twenty years earlier and now the
men mostly stayed at home. Some of my father's friends had
resisted this for as long as they could and had various cross-
country scrapes on the way back from more distant village
pubs: being chased by the police when they half-cut across
fields of turnips or barley and triumphantly tumbling into
farmhouses in the early hours covered in mud. The village
hall began to deteriorate and fill with dust. There were
fewer people in the villages, and they were getting older.
People now wanted to retire from the town to a nice village
in the countryside, and they had more money than the last
of the local farm workers, so there was a shift in the social
mix of the villages to become more middle class. There was
a cultural shift as well that came with TV, an influx of new
residents, and the technology of the modern world. Many
people's lives now seemed to revolve around the cultural orbit

of the local town—its shops, cinema, and leisure center—rather than the local farming landscape. Harvest festival and the village hall auction of fruit, jams, toffee, and bread meant something important when I was a child but had now fizzled out and become meaningless.

~

The big modernizing farmers we knew sounded a lot like Earl Butz. They had often been educated in agricultural colleges and were true believers in the gospel of efficiency. They were "businessmen," engaged in a desperate race to be the survivors, as everyone else fell behind, gave up, and stopped farming. Everything had to be big and fast. They were ruthless capitalists. Dad was a bit confused by them. He said they were "shirt and tie" farmers, "too flash" with their fancy Range Rovers. They didn't get their hands dirty and they sounded as if they worked for a corporation—listing data about their milk yield average per cow, grain moisture content, or their costs of production. They often had dozens of people working for them. The big new farms had a high staff turnover because the work was now deskilled, boring, and dirty—more like repetitive factory work than the skilled "stockmanship" or "field craft" that had gone before. Immigrant workers came and went, without anyone really knowing their names. The old farm workers had often thought of themselves as equals in work to the boss (at least in our landscape where the farms were small), but now workers didn't go anywhere near the farmhouse. Dad thought these farmers were forgetting their values, getting above themselves. Most confusing was the fact that they didn't do any of the things that made farming a joy: the hands-on working with animals and skilled fieldwork. He pitied them, even though they were a lot wealthier than he was, because he couldn't think of anything worse than being a kind of corporate boss stuck in an office.

~

My old man is sawing logs with his chainsaw. Its two-tone whine echoes back from the distant trees. I am picking up the logs—their ringed cross-sections a bright sappy orange— that drop behind him. I throw them into our trailer to take home to dry for the fire. The smaller branches are thrown into a heap and later torched. We are clearing one of the old thorn dikes that once we would have laid and maintained, but now we have given up on it.

We had mostly stopped laying hedges the old-fashioned way, by hand. Like scything thistles, it took time and men we no longer had. When the last of our farm workers, John, grew old and left, the skilled work that he had helped make possible began to go undone. At first the hedges simply became shaggier and raced for the sky, but over the years their bottoms ceased to be dense and tangled and full of different plants, and all that remained was a line of aging and scruffy thorn trees. The branches reached out and scratched our expensive tractors and a lot of grass was wasted because it went unmown in their shade. Then, with each passing storm, the old trees gradually blew over, making them no good as a boundary, so we cleared them with a chainsaw or digger. We gained larger, more efficient fields that way. Hedges and walls had become a nuisance. The hedges we didn't grub out, we flailed, using a machine attached to the side of the tractor, like a lawn mower on a telescopic arm. Soon, they were no longer true hedges with a thick, living, plaited heart, but from a distance they still looked OK, with a tidy, flat mown top. But they weren't the dense hedgerows of my childhood, and as they grew old and were lost, the landscapes became barer.

As we threw the last logs in, I looked back at what had once been a shaggy old hedgerow and remembered playing hide-and-seek there as a child with half a dozen of the village kids. And as a teenager I had seen one of the older boys shoot a rabbit with an air rifle there: it had tumbled over itself as if trying to shake the pellet from its head. Now the hedge and that past were being erased.

More than half the hedgerows in Britain disappeared between the Second World War and the present day. Thousands of miles were lost every year. Some of the hedges were many centuries old, and full of amazing and rare living things. We didn't think about that then, but it became more obvious and more important as time went on.

~

Whispering doubts about the new farming had grown in me until they were deafening. The costs side of the ledger seemed to expand endlessly as the years passed. It was perhaps fine that we weren't the winners; I could get my head around that and adapt and survive, but after a while it was hard to see anyone winning. The giant pig farm I had played on as a child was declared bankrupt and was sold up. And in place of an old patchwork landscape full of working people, diverse farm animals, and crops, with lots of farmland wildlife, a blander, barer, simpler, denatured, and unpeopled landscape had emerged.

~

The more of "progress" we saw, the less we liked it. And we always had something to measure it against, because progress somehow never quite fully happened on my grandfather's farm in the fells. We held on to that backward little farm and it became—for my father and for me—a counterpoint to the new farming. Unexpectedly, this odd combination of two different kinds of farming changed my family. We saw, on a daily basis, a "before" and "after" version of a British farm. It invited judgment and enabled comparisons, something not generally possible for other farmers. But the clearest comparison came to us from the roadside between our two farms, where an old man called Henry had lived.

~

Henry looked like the stereotypical old farmer. He was heavyset and red-faced. He wore tweed trousers, with bale string and wisps of hay peeping out of a crumpled jacket pocket. He walked with a steady gait and spoke steadily too. We didn't spend much time with him, but we knew Henry to be a good man. His rented farm was on the same estate as ours. It was a fine old place with big stone barns and a beautiful house, grander than our place. It would once have been a prosperous farm. Now it seemed outdated, lacking the giant steel-frame buildings of the modern farms.

To us, mention of Henry's name was a gentle joke about a farmer who had never modernized and had been left behind by his progressive farming neighbors. His was an old-fashioned mixed farm with fields of barley and turnips grown in a rotational cycle. In winter he fertilized his fields by spreading straw-based muck that had come out of the cattle yards and had been rotted in the midden, instead of artificial fertilizers or slurry, which we had now all begun using. Flocks of large black-faced Suffolk sheep grazed on the stubble and the turnips. Big broad-backed Hereford bullocks grazed on the pastures. Henry was a good farmer, but a generation or two past the sell-by date. He was stuck, in our eyes, in doing things the way his father had done them. He was uninterested in change, and yet somehow he survived. We speculated he got by because he didn't have a wife or family and could live "quietly" without much money.

Like the last giant tortoise, Henry seemed stranded in the past as everyone like him disappeared. When I was young, my father would comment on Henry's old-fashioned ways as we passed his land on the main road. He made meadow hay late in the summer, not silage like all his neighbors. His mowing would be two months or more later than the intensive farmers down the road, and a month later than ours. "Good old Henry is just now mowing his meadows . . ."

Dad liked Henry. He would poke fun affectionately at his "backwardness," but over the years the tone of his voice came to speak more of admiration than contempt. For a long time, I didn't listen. I would tune out, watching the lapwings rise above Henry's fields, their paddle-like wings flashing white, black, and emerald in the winter sunshine. I didn't really understand why until years later, but I can clearly remember that the sky above his farm swirled with curlews, lapwings, and rooks, and flocks of fieldfares in winter.

When Henry died, his neighbors seemed sad that one of the last of the old breed had gone. He might not have been at the cutting edge of farming, but he was well liked. The estate that owned the land carved up Henry's farm into parcels and added them to the neighboring bigger and more modern farms. One of my father's friends farmed next door and took some of what had been Henry's land. He sent in a soil analyst to work out what he should add to the soil to make the land more productive (assuming that it would need artificial fertilizer or lime to get it into full efficient production). Intensively farmed land was regularly soil-tested to see what artificial nutrients needed applying.

But the analyst reported back that the soil was some of the best he had ever tested. Henry's soil was healthy. It needed nothing. It was full of earthworms—rich and fertile. My father found this news a revelation. It shook him, because it said something about what the new farming was doing to the land. The most traditional farmer in the district had the healthiest soil. The men had discussed it in the pub. Dad said the way farming was going was insane, and that old Henry had known more than the rest of us "daft fuckers" put together.

For weeks afterward, as we passed Henry's farm, Dad would tell me that we were bloody fools. This news confirmed something Dad had always felt in his gut. Deep down he had never really believed in many of the changes, and with every passing year he was becoming more skeptical. We were doing these new things because we had to—getting more cattle and sheep, acquiring bigger machines, making these changes and meanwhile losing good people—and yet where was it all heading? If modern farming made the soil worse, and reduced it to a junkie requiring more and more hits of shop-bought chemicals, then how sustainable was it? Dad couldn't step out of it entirely, but he saw right through it. Rather than admire our friends and relatives who were creating huge new farming businesses, with enormous buildings and loads of machinery and staff, he worried for them. He thought their world was ugly, built on debt, and increasingly risky and volatile; it would all come crashing down around them someday. And when it began to, and some of the biggest farms went bankrupt, he defended them and said we had all been fools once. There was no pleasure in seeing farmers losing their farms.

My father knew the truth lay in Henry's soil.

~

The old farmers said that where there was muck there was money. They knew they had to feed the soil—and with the right stuff—or they were robbing it and would come unstuck.

The manure in my grandfather's middens was full of digested roughage from crops like hay. It rotted down over the winter and then was cast onto the fields by the muckspreader like a kind of compost. But as we shifted to feeding our cattle with silage and bought-in high-protein feeds, our farm began to produce more and more muck of an entirely different kind. Our cows now splattered jets of slurry out of their rears. Slurry is much higher in nitrogen than midden muck and emits noxious gases. It cannot be kept in a heap, but we didn't have an expensive new storage lagoon for it either. Dad was constantly worried that if we didn't spread it straightaway, it would run off the yards and down to the river and pollute it and we would go to jail. So I spent nine winters spreading shit.

The new farming had taken two mutually beneficial things—grazing animals and fertilizing fields—and separated them to create two massive industrial-scale problems in separate places. The farms with thousands of animals had more muck than their land could possibly accommodate, while the crop farms now had no animals, and thus no muck to fertilize plants, so were entirely reliant on Haber-Bosch fertilizers. Livestock in the new systems were now creating muck so acidic that the soil it was spread on began to compact and die. Crop-growing farms were top-dressing with ammonium nitrate and killing their soil. Everywhere, on both sides of this insane division, the unseen living things in soil (billions in every teaspoon of healthy soil) that had once made it all work were being killed off.

~

A mile or two past Henry's land my father pointed to a field being plowed by the roadside. A giant red tractor was pulling a huge blue plow. I could sense he was alarmed by something. "Look," he said, "there are no seagulls or crows following the plow." This was a shocking thing to him. "There must be no worms in those fields. They've all been killed off with slurry," he said.

~

In those years, my bedroom window looked out over the farmyard, and three feet beneath the sill were a series of through-stones jutting out. I had never cared much for being in the house, and had snuck out ever since I could reach the stones, and had never shaken the habit. After working long days with my father, I was sick of being told what to do and of him knowing my every move, so I would go wandering at night. I would dangle my legs out until my feet were on those stones, climb along sideways, Spider-Man-style, until I got to the back-kitchen roof, which sloped down to the field. And then I would be off and away out of anyone's sight. Five minutes later I would be over the brow of the hill. I would often go and watch our flock graze and would let the growing lambs come up to me, and judge which would be the future stars. When the sheep wandered off, I would lie with my back to a stone or tree trunk or climb an old oak tree. I loved the wild things that still lived in our fields. The hares that grazed among the sheep. The oystercatchers that laid three eggs in the top of the rotten gate stoop by the field I had worked with my grandfather years before. The rooks that foraged for worms and leatherjackets in our pastures. The kestrel that hovered above the rougher grass by the dikes. Perhaps, above all, I loved the curlews ghosting over our pastures and meadows, making their wind-blown cries. My whole life had been spent under their wings and their song.

~

One night I took an old paperback with me, scrunched into my pocket, and read with my back to an old oak tree. In our local town there was a little bookshop, and I would go there for stuff to read. The guy who ran it was a bit of an old hippie. He sorted out books about nature onto a table to greet people entering the shop. In among books about otters and peregrine falcons were some books that were angrier and political—books about ecological decline in the countryside. I had been brought up to steer clear of such stuff. My family's exposure to the early messages of ecological doom on TV led us to think that the environmentalists were bonkers. My dad despised being preached at by people who clearly had more comfortable lives than he did. After one such news report, Dad turned the TV off and declared, "If them cunts get their way, we'll be running a fucking butterfly farm."

The book in my hands under the tree was Rachel Carson's *Silent Spring*. I feared it was going to be an anti-farming rant. But instead in its pages I found a well-argued confirmation of what I had suspected for some time: the new farming technologies and practices weren't benign tools of progress. They were an arsenal of chemical and mechanical weapons that had profoundly altered the natural farmed environment by overturning its biological rules. After reading for half an hour, I lifted my eyes from the page to the fields in front of me, and I knew she was right.

Rachel Carson is known as the woman who woke the

world up to the dangers of pesticides—particularly DDT. *Silent Spring* came out in 1962 and said something remarkable—that the industrial dream of farming progress was flawed. She revealed that pesticides were poisoning whole ecosystems, and the more farmers used any chemical or medicine, the quicker they would become obsolete. It was a biological certainty that weeds, bugs, and bacteria would soon become resistant to the sprays. (Carson, an American marine scientist and conservationist, is often mistakenly believed to have been against all pesticide use, but in her book she actually argues for highly targeted use of pesticides where necessary and biological solutions wherever possible.) She saw that farming's answer (provided by corporate chemists) was always to escalate, to apply more and more, ever-stronger chemical solutions, creating a desperate race between agricultural chemists and plant and animal DNA to maintain their control. This led to even more destructive chemicals being used in delicate ecosystems that, Carson argued, human beings still barely understood. Farming was trying to break out of the cast-iron rules of biology, rather than work with natural processes. Farmers, egged on by politicians, economists, and big business, were trashing the very systems on which life on earth depended, and barely knew they were doing it until it was too late. They had become too powerful in their fields and no one had noticed what that meant for nature. Could citizens trust such farmers, chemical companies, or even the government to do the right thing?

~

That book, under that tree, changed everything for me. The landscapes falling apart (including ours) were not, as Schumpeter saw it, "creative destruction" but plain old-fashioned destruction.

I felt as if I had woken up from a long coma. I had almost conditioned myself to exclude nature from how we thought about the farm. I had begun to view my grandfather's farming with contempt, to pity my father's reluctance to modernize. And now I felt like a bloody fool, because my grandfather had been right to try to resist, and my father was right to instinctively distrust it all. I had been editing the world in my head to make it seem OK. But now the awkward memories elbowed themselves to the front of my mind.

~

The morning after I had sprayed my first field of thistles, I went down the lane to check on a robin's nest that I'd found a few days earlier. It was close to where the thistles had curled over from the chemical spray. The chicks were dead in the nest, cold bundles of pink skin and bone and scruffy feather stubs. I knew this was my fault. A tiny voice inside me had said it was wrong. I think I told myself that three or four chicks were a one-off cost to get a big problem sorted, that they might have been killed by us mowing thistles in some other way. I'm not sure I believed it, because when I remembered those dead chicks, I felt ashamed. And now, after reading *Silent Spring*, I knew we had been sleepwalking. So I began to read more and more about Carson and the critics of farming.

~

Silent Spring unleashed a storm of controversy. The corporations making and selling the sprays, and the big-farming lobby, fought back savagely, but Carson wouldn't be beaten, and DDT was eventually banned. The problem was that the campaigning focused on DDT. Her more significant insight—that farming had broken out of its old constraints—was largely ignored or forgotten.

Intoxicated with technological change, governments and farmers preferred to believe that DDT alone was a hiccup, rather than a symptom of something more troubling concerning human beings and their power over nature. And so, despite Carson's challenge, by the 1970s the brutal pursuit of industrial efficiency on farms to deliver more and cheaper food became ingrained in agricultural policy across the developed world. There was almost no acceptance in farming or in the political sphere that the insatiable pursuit of industrial efficiency on the land might itself be the problem. Instead, the processes changing farming were increasingly championed as "progress." Postwar societies were telling farmers that their job was to produce vast amounts of cheap food, and to use whatever tools were required. Many farmers wanted to hear this and embraced the changes. Others were swept along behind them in an attempt to survive. This new culture told "consumers" that food was little more than fuel, and that it should cost less and less of their income. If nature was thought about at all, it was assumed to be robust enough to manage somehow.

It was almost as if Carson had never existed. The environmental movement around the world grew massively as a result of her rallying call, but the intensification and industrialization of farming persisted in defiance of her warnings for decades after her death in 1964.

~

The light across the fields was dying. We were leading the last of that year's hay bales into the barn because the weather forecast had predicted a downpour overnight. The curlews over the brow of the field seemed agitated as if they were seeing off a crow. I looked at the fields and wondered what harm my father and I had done, and whether our land was now degraded. I knew that across the farming landscape some birds were no longer there that once had been. The corncrake had quickly vanished when farmers adopted mechanical mowers and tractors in the early twentieth century, but curlews thrived on farmland in large numbers for decades after the transition to tractors and machinery. They seemed to love the way we farmed. In spring they would appear, back from the winter mudflats to their ancient breeding grounds on our fields. They would wheel around our farm in giant fairground-ride loops, calling for their mates. Or they would walk through the grass not far from where I sat—all stilt-like legs and elegant curved beaks. They would hunt for worms on the plowing and in the pastures. And soon they would be paired up and would nest. In the weeks that followed they would fill the skies with their calls. As my grandfather had taught me, I enjoyed discovering where their nests were, and then, if I was mowing the grass for silage, I would try to ensure that the chicks escaped, even if I had to leave a strip of grass unmown for an hour or two and come back later. There were three or four pairs in our silage fields, and every summer they would rear a handful of chicks. It was hard to reconcile this abundance of birds with the reports we were starting to hear about ground-nesting farmland birds disappearing, though I'd sometimes wonder whether we had slightly fewer than we used to.

How many curlews had there been twenty years ago? No one seemed to know, when I asked. And no one was worried. There were curlews all around us. But I could sense a change when I went to help out at my friends' farms farther down the valley, on better, more "improved" and more intensively farmed land. I was aware, when we climbed off the big tractors for meals, that the skies above us were empty except for the odd passing crow or seagull.

~

One of the hardest aspects to understand in how farming affects nature is that there is often a time lag between cause and effect. We find it easy to comprehend the immediate destruction caused by a sudden catastrophic event: miles of ocean spoiled by an oil slick; the life in rivers killed by chemicals that have been dumped; elephants butchered for their tusks in the red African dust; and huge-trunked trees crashing down to the soundtrack of chainsaws in the forests. Everyone with an ounce of conscience and intelligence knows that such sudden destruction as a result of human behavior is wrong. But in reality the world doesn't fall apart on one day. It is much harder to see and appreciate the gradual changes that destroy things over ten, thirty, or a hundred years. The tools or practices that radically changed our farmed landscapes emerged decades ago. There was a considerable time lag before the science began to show devastating declines in certain species. This is because it took time for new technologies to be adopted by farmers and scaled up to use on their land, to bring about their full effect. Nature can hang on for years, even decades, before it is beaten in any particular place, and then vanishes.

~

In my father's last year or two we often spoke about what had changed and why. He was sad that the curlews, once the commonest of farmland birds, had become vanishingly rare. We both knew that it wasn't modern farming, or even making silage, that did in the curlew, but a particular level of intensity being reached in those activities. From the 1990s onward, super-big, super-fast mowers emerged and were used on fields all around us for making silage earlier and earlier in the year. The grass could be ready then because of changes to grassland management, powered by artificial fertilizers, and faster-growing grass seeds, and encouraged by the trend to feed younger grass for peak nutrition to dairy cattle. This combination of practices sent curlews into a disastrous decline. They couldn't lay eggs and rear chicks in fields that were mowed three or four times each summer, starting as early as May and ending in October. The fields where they had once lived became a killing zone for their chicks. A driver on a huge new tractor (often young, working for a contractor, with the radio on and the windows closed to keep his expensive machine clean and tidy) had no time to stop and lift a curlew chick out of the grass, even if he had spotted it and recognized it. Adult curlews tended to linger in the fields long after the changes were introduced to the way the fields were managed, because they can live up to twenty or thirty years of age and faithfully cling to old nesting grounds, trying over and over again to lay their eggs and raise their chicks.

A farmer would only know the curlews were in trouble when they were no longer there at all, as the parent birds died off. By the time he understood there was a problem, it was already too late. But was it the farmer's job to care for curlews? Or was the priority to make the best quality silage, to make milk available at prices demanded by the supermarkets? How much was a curlew worth?

The problem wasn't necessarily that mechanical or chemical farming tools existed; it was that they were being used in such a way to bring about conditions in a landscape that made it impossible for wild things to live or breed. The intensity dial was turned just a touch too far, and most of the time it was impossible for the farmer to know how to read the effects. When does the speed of a mower become a problem? When does a combine harvester become so efficient that it leaves too little grain on the field for wintering birds?

Farmers often didn't know enough to make informed ecological decisions effectively or couldn't see how they could survive financially if they made themselves less efficient than other farmers by opting out of modernization. When machines were bought, the farmer was simply updating or upgrading his gear. He didn't buy them with wildlife in mind. They weren't designed, engineered, or sold with wildlife in mind, either—the tractor designer, engineer, or salesperson hadn't a clue that what they were creating might have these unintended consequences. The supermarket that demanded food be produced in ways that required this degree of mechanical efficiency didn't have a clue either. The whole system was so fragmented and specialized that most people working within it were either ignorant of its unintended effects or, worse, lost in a kind of magical optimism that somehow nature would be OK. There were profoundly important questions about the potential effects of each new technology that it was nobody's job to ask or answer. There was no mechanism for farmers or ecologists to judge whether a technology or new farming practice was on balance a "good" thing or a "bad" thing, and we didn't really know when we had crossed the invisible threshold from one to the other.

As I moved through my adult life I witnessed hundreds of little shifts that together added up to a transformation: the gateways widened for the new, more efficient combine harvester, the slightly wider and faster mowing machine, and the bigger, deeper-cutting plow with more furrows. Slightly stronger pesticide was sprayed to protect the crops; new grain or grass seed that had been treated with some chemical or other was sown; the switch was made from spring-sown grain to winter-sown varieties that eliminated stubble where the birds used to glean in winter; slurry was spread instead of muck from a midden; and a top dressing of artificial fertilizer richer in nitrogen made the grass grow faster, in order to be mown earlier. These changes built on similar, relatively benign farming practices that had been used for centuries. The differences in these practices in my lifetime were gradual revolutions of scale, timing, uniformity, efficiency, and speed.

Over thirty or so years, the poet Virgil's farming tools for waging "war" evolved from being the battlefield equivalent of spears and swords to something more comparable to tanks, jet fighters, and chemical and nuclear weapon systems. And, in turn, a culture war that began with Carson's awakening grew increasingly polarized and toxic. On one side was a group that said farming was doing an indispensable job, becoming more efficient and making sensible use of the best new technologies, and that all was well. On the other side were those who believed that farming was trashing the earth. One encouraged a kind of naive faith that farming was "good," the other a kind of activist anger that it was "bad." One group acted as if the only thing that mattered was cheaper food, the other as if food didn't matter at all. Everyone was compelled to take sides. It grew more and more vitriolic, the arguments more and more reductive, but it was largely a dialogue of the deaf, and it existed even in our village. My first brush with this culture war had come years earlier, not long after my return from Australia.

~

My father was setting the world on fire. Silhouetted against the skyline at the top of the sandy bank above the billowing gorse bushes, towering, flickering tongues of flame and sky-bound columns of smoke rose between us. The fire licked and hissed through the tinder-dry needles. The strongest branches and trunks were as thick as my wrist and they crackled as they began to glow. The whole bank on either side of the smoke was a mass of canary-yellow flowers. Dad climbed through the gorse, setting it alight as he went. He had in his hand a kind of torch made out of a branch and an oil-soaked rag. The flames roared around him. Rabbits fled, their cotton-wool tails bobbing. A blackbird spluttered out of the undergrowth and away; small flocks of linnets and yellowhammers flittered down the field.

The gorse had been slowly creeping outward, across the Quarry field, taking over about a third of it. My grandmother had once kept hens there, and later my father kept pigs on it because they helped its sandy soil. Now it was a haven for rabbits. It was one of our worst fields and got overlooked for a few years between these periodic attempts to tame it. That night, the mood took Dad to do something about it. He marched out of the yard with a box of matches, a plastic bag full of straw and bale string, and a small jerry can of petrol. My mother sent me a few minutes later to see he didn't "burn himself to death."

After a while, the roaring, crackling firestorm began to wane and lose its wild power. As I walked back to the village between the crooked silver limestone walls I met one of our neighbors. Her husband was the head teacher at a local school, and she had a small holding of a couple of fields around her house. She seemed agitated and looked wild and emotional. She asked what had happened. Then, before I could answer, she said she had called the fire brigade. I told her there was no need. The fire was under control. My dad had lit it. It was his fire on his land. She seemed puzzled and a little angry. But she was also confused because she was a friend of ours. She was trying to understand what was happening to a hillside near her house that she clearly loved as it was—before my dad turned it to ash. She had only lived here for five years, and had no idea that this was part of a traditional cycle. She thought the way this field was now was how it had always been and how it must always be. I could see she was trying to balance her anger and her usual respect for us. She wanted me to explain.

Why is your father burning it?

Why is he destroying the birds' habitat?

What will happen to the wildlife that lives in the gorse?

Her tone said, why is your father being like this now, when usually he is such a decent man? What is wrong with you people?

I didn't have all the answers. I'm not sure she was in a mood to listen anyway. I was embarrassed. I saw that two different ways of seeing our world were clashing. My dad only rarely explained in words why he did things. Behind me on the hillside he was acting a bit heathen and wild; the wind was beginning to catch the fire in a new patch of gorse and it was roaring upward into the sky, crackling loudly.

Had Dad thought about the bird nests? I had no idea.

I felt a bit like the PR man for the Apocalypse. But I also felt a kind of defiance and a sense of loyalty to him. I wanted to defend him and explain why he would do something that seemed so wrong to her. So I tried to say that fire was how we had always dealt with the gorse, part of a very old cycle of managing it, knocking it back when it spread too far, and then letting it recover for a few years. Dad didn't burn all the gorse. He burned some patches and left others. The birds weren't entirely robbed of their habitat. The gorse is a farming nuisance; it ruins the sheep's fleeces. It would, if unchecked, take over the field completely, making it fit for little more than rabbits. I think I said other things as well, some of which even made sense, but perhaps I also said some things that were close to lies. I was powerfully aware of the widening gulf that existed between people who thought it natural and necessary to shape the landscape and those who were troubled by it; between people who farmed and those who didn't.

~

After my father's death I remembered that night of burning gorse and realized that many people who cared about nature were drawing the conclusion that all farming and all land management was now to be distrusted. But to reject farming wholesale is a mistake: the field is the base layer on which our entire civilization is built. We all depend for our survival on those bits of wild land that have been cleared by means of fire or using an ax, or by grazing animals over time, or by using a plow. The fact that this clearance probably occurred so long ago means we forget this harsh ecological truth and almost think of fields as a natural phenomenon. But a field is not a natural thing (though some fields, like species-rich hay meadows, or carefully grazed prairie, or woodland pastures, can closely resemble wild habitats), and whether it is used to grow plants or animals, it ultimately comes into being as a result of killing some of the original species. Many of the creatures that would have lived in that space now cannot, as they are excluded and no longer have spaces and food that would sustain them. The creation and sustaining of a field can mean life for some things and death for others, and always did.

The truth is, we are quite brutal ecosystem engineers and change the world for our own use. Once, we cleared our own fields in order to grow food, and had to harden our hearts and do this landscape engineering to produce the food to keep ourselves alive (as most people in the developing world still do). By the time I was an adult, barely any British people still worked on the land (less than 2 percent of the population), so collectively we had begun to forget or turn away from these difficult truths—just as the head teacher's wife had done.

The enduring truth is that we all are—and will always be—complicit (directly or indirectly) in killing for our food, regardless of what we eat. It was a battle long before Virgil said it was. Farmers had to fight to come out on top of the countless things that would destroy their crops or kill their animals: they have always been killers, and whether they grew plants or kept animals made little difference to this reality. Farmers destroyed wild habitats to carve out fields for crops, just like my father did that night (and, in doing so, created new habitats and new ecosystems for other species). If you want to put this to the test, try living from a piece of wild land and see how long you can avoid killing anything (and don't forget to count the small living things, because otherwise it is cheating).

~

The logic chain is simple: we have to farm to eat, and we have to kill (or displace life, which amounts to the same thing) to farm. Being human is a rough business. But there was a difference between the toughness all farming required and the industrial "total war" on nature that had been unleashed in my lifetime. Despite their inevitable ruthlessness, the original farming societies often had ethical or moral codes that forbade overexploitation of the land and emphasized the need to take nature as a guide. In the Old Testament, Leviticus 19, God gives Moses the Commandments that the Jews should live by, and in among those well-known rules, most of which form the basis of our own laws, is this forgotten and long-ignored instruction:

> When you reap the harvest of your land, you shall not reap to the very edges of your field, or gather the gleanings of your harvest . . . you shall leave them for the poor and the alien . . .

~

My father wasn't much of a churchgoer, but he believed in something similar. He thought that things should have limits and constraints. He believed in moderation and balance. And he died hating what had happened to farming. He had seen enough of it to know it had become a corruption of all he had loved and cared about. In the last decade of his life he had no time for its fucked-up logic. He was saddened by what it had done to families, rural communities, animals, and nature. He lost interest in trying to keep up with the big farmers on our own land. He acted as if it was all a stupid game that he wasn't interested in playing. So he looked after his own land, and just held on. He never got to see the new farming at its global extreme and wouldn't have thanked you for the opportunity. But in the months after his death, my wife, Helen, and I traveled to the American Midwest for the first time. Twenty years earlier I had seen something of the new farming emerging in Australia (enough to know what was coming), but I knew now that the Midwest was the logical conclusion of all that was happening. The efficiency endgame. My farming apprenticeship ended by seeing that future in its purest form.

~

We drove down the highway, past shabby farmsteads with flaking paintwork and rotting wood, past tumbling-down tobacco barns cut through with shards of sunlight. Past abandoned cars and rusting farm machinery, and black cattle standing in paddocks next to farmhouses. Past towns that seemed half-abandoned, with boarded-up shops and houses with Confederate flags in the windows and VOTE TRUMP signs on the front lawn. Shutters were closed and leaves gathered on the porch; churches with billboards promised redemption for drug addicts. Flakes of snow fell but didn't settle.

We had traveled to the heart of American farming country to stay with an old friend in Kentucky. It was winter and it felt like it might never end. We were made welcome in the white clapboard farmhouse that was full of books. We ate good simple food and talked about our families and our farms. But as hard as we tried to be cheerful, it felt as if we had stumbled into someone else's grief. There was a sense of impending doom about the coming election results. This had once been a thriving landscape of small- and medium-sized farms. Now it felt like a landscape littered with ghosts and relics.

Our friend drove us around the county in his white pickup truck, with his sheepdog in the back and his red toolbox and wrenches in the footwell. He told us about his people, past and present, and introduced us to farmers who were holding on. They all told us the same thing: America had chosen industrial farming and abandoned its small family farms, and this was the result—a landscape and a community that were falling apart. They showed us fields of oilseed rape that were full of weeds because they were now resistant to the pesticides that had been overused. They spoke of mountains ripped open for minerals, and rivers polluted, and farming people leaving the land or holding on in hidden poverty. And the worse it all got, the more people seemed to gravitate to charlatans with their grand promises and ready-made scapegoats to focus all their anger on. I felt I had landed in a future that didn't work, and the people I met sensed my unease. "You haven't seen anything yet," they told me.

~

The vast black fields of Iowa go on forever. The soil, rich and deep, is flecked with the stubs of cornstalks. It lies exposed to the wind and rain for half the year. And grows like hell the other half. I heard the young woman say that she loved this landscape, that in summer you could "hear the corn growing." But to my old-world eyes, this wintry desert had little romance or history in it.

It is a landscape of big skies—and below, all is dark, flat, and bleak. It offers little but utility. The farms look like something out of that Grant Wood painting—*American Gothic*. And the iconic homesteader and his wife must have left for the city or are in the house watching TV, because there aren't many people in these landscapes. Everything old was rotting. Barns leaned away from the wind, roofs half torn off. Corn towers and grain elevators broke the flat black horizon and shone silver in the sunlight. Giant pyramids of orange corn stood in the rain, under the arms of grain elevators. The plowed fields butted up to the picket fences of the crumbling farmsteads and stretched from horizon to horizon. It was an expanse of corn, soybeans, and pig sheds. This was the agricultural landscape that Earl Butz demanded.

I was traveling with an agronomist whose passion was soil, and how to change farming to protect it. The troubles of this landscape hurt her, because this was her home and the farmers here were her people. Judging these places harshly is easy if the farmers aren't your family and you don't see the bags under their eyes and the stress on their shoulders at family parties. She told me this landscape was created in the supermarkets of America—by the cult of cheap food. The people in those shops seemed not to know, or care much, about how unsustainable their food production is. The share of the average American citizen's income spent on food has declined from about 22 percent in 1950 to about 6.4 percent today. But it is worse than that, because the proportion of every dollar spent on food that goes to the farmer has declined massively to around 15 cents and is still declining. The money that people think they are spending on food from farms almost all goes to those who process the food, and to the wholesalers and retailers. The winners are a handful of vast corporations that have politicians and lawmakers of all political parties in their pockets.

The agronomist told me that Iowa is blowing away. For

half the year, the wind across the tilled fields slowly steals the topsoil, carrying it to someplace else, one tiny particle at a time. It doesn't seem much on any given day, but it is relentless and in places several feet of topsoil have been lost over the past century: an endless and unsustainable waste of the soil that feeds America. The reality can be seen on the dirty brown snowdrifts, the stolen wealth of this land caught, for an icy moment, before being lost.

If this was the future, it was curiously shabby and ugly. It didn't appear to be very nice for people to live in. Many of these farms, even the prosperous-looking places, were deep in debt. The landscape was an ecological disaster, as sterile as could be, and was responsible for a dead zone in the Gulf of Mexico, where the eroded soil and field chemicals all flowed once they got into the Mississippi River. A lot of the work was done by Mexican immigrants, who had been displaced from their own farms by the American corporations bullying them out of business. And what couldn't be done by cheap immigrant workers was done by machines, which were now self-navigating, and able to do the work in the field guided by satellites. These farms could impose their will on the land as never before, and increasingly the "farmer" didn't even need to be there to do it.

We parked by a ghostly farmstead, dwarfed by the pig confinement sheds and the silver grain silos beside them. Suddenly, out of the trees to our left, a large, black, muscular shadow turned with each wingbeat into a bald eagle, the emblem of the United States. It wheeled overhead and my heart beat in my chest. The agronomist told me that the eagles had returned in recent years. "Oh good," I said. "What do they eat?" There was an embarrassed silence. "Maybe the dead pigs, outside the confinements," my host replied. We both fell silent as the eagle flapped away across the fields.

~

There are no winners here. The farming businesses that rule these fields have got so big they are entirely reliant on one or two monopolistic buyers who screw them on prices and can bankrupt them at will. The money flows off the land to the banks that finance the debt on which it is all built, to the engineering companies selling the tractors and machinery, the synthetic fertilizer and pesticide corporations, the seed companies, and the insurance agents. And yet, judged solely as productive businesses, focusing on efficiency and productivity (and ignoring the fossil fuel inputs and ecological degradation), these new farmers are amazing—the best farmers who have ever lived. In the year 2000, the average American farmer produced twelve times as much per hour as his grandfather did in 1950. And this amazing efficiency means the end for most farmers. In the UK, the number of dairy farmers has more than halved, from more than 30,000 in 1995 to about 12,000 today. In turn, the number of dairy cows in Britain has halved in the past twenty years. The amazing productivity of the remaining farmers and super-cows is demonstrated in the simple fact that milk production has remained more or less stable.

Statistics like these have resulted in transformed lives, changed diets and household budgets, and completely different ways of life around the world. The share of British household income spent on food and drink has declined from about 35 percent of the average household budget in the 1950s to about 10 percent today (though poorer people spend a greater share of their income on food and drink, say, 15 percent). The money we once spent on food has been freed up to spend on housing; leisure activities; consumer goods like cars, mobile phones, clothes, books, and computers; or on things like mortgages or rent; and on foreign holidays that few could afford a generation or two ago. The modern world has evolved out of these gains. But how people lived and shopped was at the same time creating immense pressures that originated in fields in which farmers were forced to search for every productivity gain possible. There was a direct link between the availability of cheap food and farmers having to adopt industrial techniques to work their land. And the more industrial it got, the less involved in it most of us became.

Where once your grandmother would buy a chicken from the local butcher or farmer at a market, and could ask them to vouch for how it had been raised, now we buy anonymous meat that is already chopped up, deboned, and encased in plastic. The reality of the chicken's life and death is hidden from us as if we were children. Most of us don't think about a carcass, or about making a range of meals from all of its parts. We don't make broth from the bones. We don't know how to perform the basic tasks that make eating it possible: the killing, the plucking, and the chopping. Nor can we influence how the chicken is farmed. The food company is so big it is unlikely to notice our concerns as we stand in front of the shelf at the supermarket.

Local food production once gave people greater ability to see and judge farming, and to influence it. In those face-to-face transactions between our grandmothers and the farmer or butcher, the buyers could tell the man with the chickens what they liked or didn't like. The local food market was more than a financial or commodity exchange, it was also an exchange of knowledge and values; it bound people into a kind of shared morality about how things should be done. The supermarket food system told us we didn't need to worry about such things anymore. We are all encouraged to be apathetic and disinterested. Nearly all the food scandals and farming crises that have eroded our trust in farming have come about as a result of the drive to reduce costs—through dubious practices, as farmers and others in the food chain have sought to cut corners behind the scenes, doing things their fathers and grandfathers would never have dreamed of in order to make food cheaper than it really should be.

This was business-school thinking applied to the land, with issues of ethics and nature shunted off to the margins of consciousness. There was no room for sentiment, culture, or tradition, no understanding of natural constraints or costs. The modern farming mindset didn't recognize these external things as relevant. This was farming reduced to a financial and engineering challenge, rather than being understood as a biological activity. It was exactly what supermarkets demanded, because it could guarantee year-round supplies of food products, with entirely uniform product consistency. We convinced ourselves that farming was just another business, subject to the same rules as any other—but that is coming to seem like the most foolish idea ever.

We created a society obsessed with food choices and ethics, while disconnecting most people from the practical agricultural and ecological knowledge to make those choices. Now people worry about *what* they should eat but have largely lost sight of *how* their local landscapes should be farmed and what foodstuffs they can produce sustainably. Most people are now largely illiterate when it comes to agriculture and ecology. This is a cultural disaster, because the global challenge of how we live sustainably on this planet is really a local challenge. How can we farm in ways that will endure and do the least harm? And what does that local farming produce for us to eat? This is not an argument for entirely eating local foodstuffs—I like bananas as much as the next person—but a reminder that it is good sense for a lot of our food to be produced around us and under our gaze.

~

The economists are wrong. Farming is a business unlike any other because, crucially, it takes place in a natural setting and affects the natural world directly and profoundly. As farmers collectively made the land more efficient and sterile, whole species of birds, insects, and mammals vanished and ecosystems collapsed. All the major surveys of wildlife in Britain in recent years have revealed the same thing: an astonishing and rapid loss of wild things from farmland. One comprehensive survey spoke of Britain being "among the most nature-depleted countries on earth."

From an ecological perspective the effects of the changes in farming are black and white, because we judge the outcomes, rather than engaging with the causes and processes. The human causes, as my family experienced them, were complex and confusing and difficult to navigate. It would have been much simpler if I could have picked the good guys and the bad guys and told a morality tale. But the truth is messier and more nuanced than that. Ethically it is complicated. My family and my friends did these things: good people, not fools or vandals. The financial pressures on them were and are immense. The levels of stress and hefty workloads do not provide the right conditions for seeing and valuing nature, or for enlightened thought. Some of the criticisms target the remaining farmers, most of whom are going broke trying to hold on to their land and way of life, instead of the large corporate and political forces. They are seen as the ones really driving farming in the wrong direction. But this is the equivalent of blaming the badly paid checkout workers in the supermarkets for the sins of the billion-dollar companies they work for.

~

None of this is to say the old ways were perfect: they weren't. Some of the changes were good and necessary, and perhaps inevitable. And it wasn't an absolute shift from perfectly bucolic to hellishly industrial. Some farmers still try valiantly to show the same respect and love for their land as their families always have. Some of the new farms have improved animal health and reduced rates of disease; not all animals in the old traditional farms were well treated, and not all animals in the new farms are badly treated. Many of the changes on the new farms are actually evolutions of older practices, like selective breeding. In practice, farms in Britain and elsewhere, including ours, are all now somewhere on a spectrum between the most intensive industrial farming and traditional farming. Almost all have modernized, but some much more than others. "Farming" is now a term that tries to encompass a vast, messy range of activities.

And so now, as we search desperately for a way to make things right, we must avoid overly simplistic solutions. History gives us many examples of reductionist thinking being applied to farming with the result that things just get worse. Miracle fixes often come with unforeseen consequences.

~

In 1915, German government officials were worried about the lack of food available for their population because of Allied wartime blockades. They decided that slaughtering Germany's 5 million pigs was a good idea. The pigs, they reasoned, were consuming too many plant calories that could otherwise be used for people. The slaughter was called the *Schweinmord*, or "pig massacre." It seemed to make perfect sense. But the officials had not understood the role of the pigs in providing manure on farms to fertilize the crops, or that many of the pigs were eating waste products from other crops and households, and therefore converting inedible or wasted organic matter into highly nutritious food for humans. In practice it was a disaster. The following year crop yields fell, and food was even scarcer than before.

~

A lot of people have fallen for the corporate propaganda of super-industrial farming. Particularly, its claim that it would supposedly solve all problems by sweeping away the old, less efficient farming, dialing up the intensity of the most industrial farms and somehow freeing up land to be given over to wilderness. There is a big problem with this plan. The new super-industrial farming isn't sustainable—it is the most destructive farming on earth in both climatic and ecological terms. It looks very much as if it won't endure long, simply because it squanders natural resources and destroys and wastes the soil.

~

According to the United Nations, there are still 2 billion farmers on earth. The overwhelming majority are not high-tech, large-scale, or highly mechanized. About 80 percent of people on earth are still fed by these small farmers, and their crops and livestock are vital in sustaining our farming into the future, super-industrial or otherwise.

It has long been recognized, thanks to writers like Jane Jacobs, that cities and towns that are too specialized—too monocultural, too modern, or too fossil-fuel reliant—work less well in practice than diverse places with a mix of old and new. The new and the old cross-fertilize in all sorts of unforeseen ways. The same is true of agriculture. Rather than making the old farming and crop species and farm animal varieties obsolete, the most intensive farming needs them for its very survival. The most intensive agriculture on earth relies heavily on the genetic diversity provided by the older farming systems, breeds, and crops. Far from being anachronistic and obsolete, the world's most "backward" farmers are a vital resource pool for the future.

And much of the diverse agriculture that remains is to

be found at the margins—in the mountains, remote places, deserts, forests, and jungles of the world; in places where isolation, poverty, lack of development, type of climate, altitude, latitude, soil type, disease risks, or duration of the growing season mean that intensive agriculture doesn't dominate and hasn't swept away traditional farming practices. Such landscapes are full of special varieties of domesticated plants or heritage breeds of domestic animals. This is because for more than ten thousand years human beings have had to try out different methods in all kinds of places through trial and error. In the global "library" of farming there can be found a whole range of solutions to millions of different local challenges and problems.

Giant industrial agricultural companies are crawling over these historic farmlands trying to identify, own, and patent the riches in them. When the most productive varieties of grain or corn cannot cope with new strains of disease or changes to climate, agronomists will look for the solution in the diversity of heritage grains and corns in the few historic farming systems that survive. When the fastest-growing pigs in intensive housed systems cannot cope with a disease, or some form of stress, agronomists will look for the desired robustness, vigor, and disease-resistance in the library of heritage pigs either in the wild or in historic farming systems. Our descendants' ability to feed themselves many centuries from now may well be decided by whether some now obscure grain or pea variety survives, or whether an archaic breed of tiny cow, hairy cold-climate pig, or heat-resistant chicken still exists in a muddy farmyard on a hillside in some "backward" place somewhere you and I will never go. If that scruffy chicken or pig carries with it a special piece of the genetic jigsaw, it may be vital.

The essential thing we know about the future—whether we are looking at economics, climate, or biology—is that it is unpredictable, so we need to maintain our library of agricultural diversity, for things we know now that we need, but also for many things that we do not yet know we may require in the future. The strength of diversity is that it gives us resilience and robustness for the future. It gives us options. It spreads our risks.

The critical thing is that agricultural diversity doesn't really survive in a laboratory or in a test tube. Much of the DNA and knowledge that might be essential for our future is systems-based—it survives by being used by people. Breeds of farm animals need sufficiently large populations, with lots of different herds and flocks, and viable local economies, to sustain their genetic diversity and stay healthy and vigorous. Heritage crops need to be cultivated and grown in all their varieties both to keep the old varieties alive and to develop new ones. This means that the farms and farming systems that use these plants and animals need somehow to survive and prosper, and trade, and sustain themselves economically. It is also not appreciated that the most intensive farming systems often use the older systems to produce their raw materials or breeding animals. Lowland sheep farmers in Europe often buy the crossbred daughters of the mountain sheep flocks or cattle herds for their breeding replacements, rather than wasting their most productive land to breed their own.

We cannot have only the most intensive farming and sweep everything else away. The productivity of these intensive farming systems masks their reliance on the older, more traditional systems; they need them for long-term sustainability. This can only be addressed by protecting the old systems.

Britain is still one of the richest places on earth for agricultural diversity, a result of its unique farming history and its gift of being one of the best places on earth to grow grass for livestock. Our islands are the birthplace of many hundreds of breeds of cattle, sheep, pigs, goats, horses, and ponies—most of which have clung on around the edges of the new farming on small holdings and old-fashioned farms in remote areas, and as a result of their stubborn champions in the farming community. As large areas of the world have been converted to modern livestock farming over the last two or three centuries, many of the animals farmed in those lands are either pure British breeds or hybrids of them. Hereford and Aberdeen Angus cattle, Suffolk and Leicester sheep, and Large White pigs are farmed in places as far afield as Texas, Western Australia, South Africa, and Ukraine. In Britain, farmers' lives often still revolve around the historic systems based on those different breeds. The knowledge of how to farm them and the skills to do so are vital.

The world has already made much use of our "library" of agricultural diversity to feed hundreds of millions of people over the past few centuries. Now is not the time to lose the old breeds and plant species. We should think holistically because we need to produce lots of good food in sustainable ways; we need agricultural diversity and we need lots of nature.

~

We can't dispense with efficiency or technological change altogether, that would be silly: there are way too many of us. We are heading toward a world of more than 10 billion souls by 2100, and if every farm is inefficient, then we as a species will need to utilize more of the planet to feed everyone, which will leave little or no room for wilderness and wild nature to thrive. The reason so many ecologists are seduced by agricultural efficiency gains is because it is a way for us to get what we need from a smaller proportion of land. The complicated truth is that we do need efficiency gains, and we can't live completely in the past. It is about finding a balance and being aware that other things matter as well as efficiency. We have to think about the problem clearly and holistically—not one fashionable issue at a time, as tends to happen. And time is running out: we have to act now.

~

We came back from the Midwest to the worst winter I have ever known. It dragged on and on, with gales and storms, trees felled by the wind, rain, sleet, and snow. The land was sodden for weeks, with mud and puddles everywhere. And after that came the worst storm of all: fourteen inches of rain fell on the already saturated ground in one day. The lakes were already full, the ground soaked like a sponge, and then the worst rainfall ever recorded in our valleys. The storm roared over the houses in giant, violent, battering waves; we felt like crabs hunkering down in a seashell at the bottom of the sea. The next morning, the hillsides ran with rivers where there had never been rivers before.

Over the fells from us, the rocky basin below Helvellyn became one vast bowl that was channeling water down toward the village. The water gathered force, formed a vast muddy torrent tumbling over the rocks; it gathered speed to become ever more violent, picking up stones and boulders and thundering down the fellside, scouring out the riverbanks. Not finding space in the gravel-filled channel through the village, it tore out the walls that usually held the beck, spilling over the banks. It cut new channels through the roads and tore up the tarmac, throwing it across the streets. It flowed under front doors and into shops, hotels, and kitchens. As people retreated upstairs, the river roared its way through the village and on down to the lake. With every hour it carried thousands of tons of gravel, boulders, and rock and piled it on the fields by the lake and throughout the village.

This torrent, and dozens of others, tumbled down the hillsides and into the frothing brown soup that was the Ullswater lake. It rose and buried roads and farmhouse kitchens, and fields and fences, and ripped up trees and carried them away. Just over the fell from us the lake rose ever higher, and out of it flowed a river that grew in power with every passing minute. Flood warnings began to sound downstream and to flash on computer screens across half of northern England. And still the rain kept hammering down.

People began to fear for the old Pooley Bridge, with its arches, and they stopped cars passing over it. The brown tide of pressure built and built, and then the old bridge crumbled and was swept away as if it had been made of biscuit. Downstream the water began to rise into fields that were usually safe, rising around the cattle and sheep. Farmers tried to move their livestock out of the danger areas. Friends of ours went with their sheepdogs to gather their flocks away from the muddy torrents. The sheep stood up on a strip of slightly higher ground surrounded by water. A shepherd knows in his stomach how this goes: five times out of ten they will act sensibly and can be helped to safety; but five times out of ten they will get spooked and dive into the water and be swept away. Life and death reduced to a simple deadly gamble. Shepherds can easily be drowned too, and dogs. Some shepherds will leave their sheep to use their own sense, hoping they survive the worst on the higher ground. Others cannot help but try to intervene.

Our friends sent their most experienced dog beneath the sheep to nudge them toward the fence where they could grab them. They moved slowly away from the deeper water to the makeshift pen created by the fence—just in time, because the water was already covering the grass where the sheep had stood moments earlier. And then one wilder sheep dived out of the pen into the deep water and, before the farmer could stop them, they all followed and were swept away downriver, bobbing one after another on the current. Some of them were found a few days later when the flood receded, miles downstream, their bellies swollen. Some were hanging high up in the branches of trees as the water receded. Most weren't found; others were crushed in their barns by mudslides or drowned.

The groundsman of a golf course found a cow wandering the fairway. Black and white, and beautifully clean, it had been washed downstream during the night by the furious river and spat out on the golf course like Jonah in the illustrated book of biblical stories I had had as a child. The farmer who owned it couldn't believe it was his when the golf club called him because it had traveled fifteen miles down the torrent.

Our landscape was left battered: there were high-tide

marks of rubbish across the fields and flotsam and jetsam everywhere. Even so, one person's loss could be another's gain. One farmer we knew had benefited from a smaller flood the month before when hundreds of tons of good firewood had swept down into his fields. He had bragged to his friends that it was "finders keepers." And then this later flood swept it all away again during the night, until not a stick remained on his land. His friends laughed and told him, "What comes around, goes around."

As the water receded there was a sense of confusion and shock about what had just happened. There is a bridge in Carlisle over the River Eden with a mark on it of the highest water levels for a century and a half. Not once in the following years had the water level risen higher than that mark—until the past five years, when the record has been broken twice: the first time by half a meter, the second time by a meter. No one here had ever seen rain like that before. It caused hundreds of millions of pounds of damage and ruined many thousands of homes in the floodplains. In the city hundreds of houses were flooded. When the water receded, people turned their houses inside out—furniture and possessions were thrown out in the streets in sad heaps, and then into builders' dumpsters. Filth stained the walls.

In the weeks and months after the flood there was a new focus on how land was managed upstream of Carlisle, even many miles upstream in our valley. The farmers I knew seemed bewildered by how much rain had fallen. River systems that had seemed robust now appeared fragile and inadequate to cope with what was being thrown at them. And they say this will all get worse, much worse.

~

Farmers used to assume that nature could adapt and cope with whatever we did on the land, but that is no longer credible. Our power to beat up Mother Nature has grown exponentially in my lifetime, wearing the mask of progress. And because of that we have destroyed things on a scale our ancestors would scarcely have believed. The old faith that the natural world has limitless reserves and resources has been tested to destruction. The idea that nature was vulnerable seemed like hippie or communist propaganda to my grandfather and even to my father's generations of farmers. But it has now proved to be true: nature is finite and breakable.

Our society is right to be concerned about agriculture. The science of what has happened is chilling, and the fact that the loss of nature is escalating is even more terrifying. We are burning down forests, filling the atmosphere with greenhouse gases, polluting the seas with plastic, and killing off species at remarkable and devastating rates. Being aware of these sad truths marks the difference between the era my grandfather farmed in and my own. As farmers we now have to reconcile the need to produce more food than any other generation in history with the necessity to do that sustainably and in ways that allow nature to survive alongside us. We need to bring the two clashing ideologies about farming together to make it as sustainable and as biodiverse as it can be.

~

We thought we had left behind the time when we needed to know about mundane things like animal muck, growing crops, feeding cattle, and making hay. And because we didn't value these things, and the old ways, we turned our backs on things that worked and forgot the vital knowledge and skills we once possessed. We didn't value the patchwork landscape of family farms with mixed habitats and rotation of crops and livestock, and so we allowed the countryside to become more "efficient," "monocultural," and "sterile." We didn't value the hay meadows full of wildflowers and insects and birds, and so let them disappear too. We didn't value the living soil beneath our feet, and so we let it compact and erode and cease to thrive. We didn't value the hedges and coppiced woodland, and so we turned away when they were ripped out. We didn't value simple things like cattle grazing in fields, pigs wallowing in muck, and chickens pecking in the farmyard dust, and so they ended up in vast industrial complexes. We didn't value muck that rolled firm from the backsides of cattle, that was good for the land, and so we never mourned when their feed was changed to silage and their muck to acidic slurry that kills soil.

We worshiped machines, so we thought it was a good thing when they got bigger and could reshape the fields and spill scarcely a grain of barley on the stubble. We cared so little about how crops were grown, as long as our bread was cheap, that we ignored them being doused in poisons. We didn't even notice or much care when those crops were planted in the autumn instead of in the spring and created sterile green fields in which birds could no longer feed in winter.

We didn't think it was our job to know, or care. We were too busy doing other things. And if giant corporations would give us the things we wanted (and tell us sweet little lies about how they did it), we let them. But it was an illusion, an industrial arrogance, a future that didn't work: a dystopia. Only now are we slowly awaking from this comfortable coma to realize that we are a long way from the fields that feed us and from knowing enough to make good choices. What we do know in our hearts—even the most optimistic of us—is that finding our way back will take time and faith, and radical structural changes in our relationship with food and farming.

~

My inheritance is an ancient one, the chance to live and work on a piece of much-loved land. From the moment I looked at those familiar fields drawn on those old title deeds in the solicitor's office, I knew it was both a blessing and a privilege, and a grim financial and work challenge. I knew we could live in an amazing place surrounded by nature and beauty, but I also felt a deep familial obligation (I'm not supposed to sell it or betray my father's faith in me). I knew I was now married to this land for better and for worse.

In the most practical and real way imaginable, I began to think about how we shaped the earth. I would have to work out how we could create a farm that would keep us, and regenerate our land and its ecosystems as best we could. I knew that we needed to be honest about the past and the present, and to use some imagination and courage to think about the future. I also knew that while respecting my father's work and knowledge, I would be free to change things if we could find ways to do so.

But how? What sort of future could we shape for our children? I was already determined that I would not intensify and scale up, take huge financial risks, or make factories out of our fields. But nor could I see how to manage our land entirely for nature, producing less, without going broke. I knew that if we farmed in more sustainable ways—and no one wanted to pay us to do that—then we would just go bankrupt. The applause of middle-class people who "care about the environment" isn't worth much when the bank manager says no. I knew we would have to chart a difficult and compromised course: being good enough farmers to pay our bills, while continuing to steward the wild things on our land so that we could hold our heads up and look our children, and perhaps someday our grandchildren, in the face.

UTOPIA

After more than thirty years I have at last arrived at the
candor necessary to stand on this part of the earth that is
so full of my own history and so much damaged by it,
and ask: What is this place? What is in it? What is its
nature? How should men live in it? What must I do?

I have not found the answers, though I believe that
in partial and fragmentary ways they have begun to
come to me.

Wendell Berry, *A Native Hill* (1968–69)

I think it is fatal to specialize. And all kinds of things
show us that, and that the more diverse we are in what
we can do, the better.

Jane Jacobs, interview with James Howard Kunstler
(March 2001), quoted in *Jane Jacobs: The Last Interview and
Other Conversations* (2001)

The old men sit around on the "posh" seats and a few extra garden chairs brought in to accommodate guests. They are holding beer cans in one hand and glasses half-full in the other. The dining room in these old farmhouses is often still "saved" for funerals, christenings, and gatherings like this one. A little porcelain kingfisher perched on the mantelpiece is ready to dive for fish among the swirling patterned carpet. Rain lashes on the gray window.

The farmhouse is immaculately clean and tidy, fretted over for weeks no doubt, and there is enough good food to feed everyone twice over. The house-proud women of the family have filled the tables with homemade bread, sausage rolls, sandwiches, chutneys and salads, slices of home-cooked meats, and puddings. They fuss around us like hens, clucking about whether we have eaten enough or had a cup of tea. The younger daughter of the family (who left years ago for a life in the modern world that doesn't work like this anymore) looks to me as if she is enduring a day trip back to the 1970s. She seems to be heroically putting up with all this old-fashioned domesticity out of respect and love for her parents. I catch her eye, and she smiles nervously, as if to say, "Oh heck, was it that obvious what I was thinking?" And I smile back to say it's OK.

We are at the seventieth birthday party of one of my father's best friends, David, a small man with bandy legs and a barrel chest, which makes him appear as if he is bursting with strength that his body can't contain. He is a dynamo, always buzzing with a new idea and a plan.

It is a very Cumbrian party. I am surrounded by my parents' friends and I have that sense of being not entirely a grown-up. But these are my people, for better or worse. The men all come to talk to me by turns to hear my news and to tell me theirs. Others give me a broad smile from across the room. When I was young, I used to think I was being judged in these conversations and that these men might think me weak and foolish. Maybe they did. But I see now that the stories we tell each other here bind us tightly together like the hawk-nervous chatter of sparrows in a thornbush.

One of the old men tells me quietly that he is proud of me, and I look to see if he is joking, but he isn't. I hadn't expected that. David makes his way over. I have known him since I was a small boy. I've always liked him, but I have only properly seen how fine and kind a man he is in the last few years. He sat with me at my father's wake and shared a drink. He told me that he had been my dad's friend his whole life, one of his best friends. We shed a tear and shared a smile together, and he told me a funny story about my twenty-year-old dad getting into trouble with a girlfriend's father because he had brought her home late. Then there had been a commotion in the barn, and he had ended up calving a cow with her father in a byre by the house. David told me that now that my father was dead he would try to be as good a friend to me as he had been to him.

It sounded like an earnest promise—and he has kept it. He comes to my farm two or three times a year and quietly offers advice, and he is there whenever I have questions I would once have asked my father. He buys tups (rams) from me, as he did from Dad, and is there at the sales to ensure my sheep make honest prices. I am forty-five years old and I am still being watched out for. David doesn't need to do this. I hadn't expected it and am touched by it. He has taken me under his wing, and he thinks nothing of it; this is just the man he is. Today he tells me he wishes my father were there with him to celebrate his birthday. Me too, I say.

They grew up together, clipped sheep together for cash, went to dances, chased girls, drove cars too fast, got out of scrapes, and lived alongside each other as they grew into men and started their families. They saw each other struggle, borrow too much money to buy land and somehow pay it back, and eventually achieve the things they set out to do. Some years ago, David was gravely ill. My father visited him in hospital and came back shaken. He thought he had said goodbye for the last time. But David recovered. And now he looks after everyone in the room, fetching drinks and taking time to talk to everyone, including my children. He asks them about their sheep, and their school, and sends his grown-up daughter for farm toys for my son Isaac. A box appears that would once have been his son's, and Isaac plays with the one-legged cows and three-wheeled tractors and paint-chipped metal pig huts.

The old men watch the children play. They pull their chairs closer in a circle and begin to talk. They think farming has gone wrong, badly wrong. I listen. One of them says we need to get back to farming a mixed landscape of different crops, with cattle, sheep, and pigs. "This just-farming-one-thing is no good," he says. "You are vulnerable if you farm just one thing, vulnerable to these insane swings in the prices of things. They tell you to get more cows, and want to lend you heaps of money, but when it goes wrong the bank comes for your farm and there is no sentiment. Don't trust them," he says.

Another man, whom I recognize but can't name, says even the good farming families are finished now. The interest rates rise, or the price of pork or milk falls, or the price of feed rises—and the numbers kill them. We have all been to the farm sales and stood in the rain while the auctioneers sell off the farmer's machinery and animals. The farmer puts on a brave face, standing by the auctioneer, and his wife checks he is OK by clutching his arm and smiling a message of love. Their friends surround them in their hundreds, buying things they don't really need just to show their respect and get their names written in the sales ledger.

We all know farmers suffering with mental-health problems. We all know farmers who have killed themselves.

Another of my father's best friends, Gerald, tells us that he stood on his highest field and looked across the Eden Valley and realized recently that it is just a "green desert." He says there aren't birds like there used to be. And where are the hedgehogs? Where are the butterflies? The other men look uneasy. I am shocked to hear him talk like this. He sounds a bit like the environmentalists we once all hated.

They talk about a field they all know that once grew deep, rich, golden crops of grain that shimmered in the wind and the sun. Now that field is worn out: it has the worst and sickliest crop on it that they have ever known. It has been used for grain for fifteen straight years and now needs to rest and get its soil right again with grass, and some livestock on it to replenish it with their muck. They talk about fields around the district as if they are old friends—and those old friends are not being cared for properly.

Then comes the voice of another man—whom I don't know—sitting in an old armchair in the corner. He says everything is being driven by money, the greed of big business and a few farmers who push for more and more and undermine the rest. We are bloody fools, he says, we just end up chasing each other down to the bottom. If we make money with a hundred cows, we want another fifty to make more, and, if we are losing money, we want another fifty to get out of the hole. Either way, it is always more, more. David's son turns to me and says, "It just isn't fun anymore."

These men were always fairly conservative with a small "c." I know because I had listened to them when they were younger and I was a child. But something has changed. There was always sense in what they had to say. Now there is radicalism too. And they are right, although I am shocked that they see it the same way I do. I thought I had grown away from them, but the truth is we have all grown away from the same bad ideas. We are all grasping to understand what has happened and how we might climb out of it. Then, because we are all a little suspicious of such serious political talk, we spontaneously get up to go for pudding, and shuffle into the kitchen like penguins in a queue. Sherry trifle with pouring cream. Lemon cheesecake. Finely chopped fruit salad. And then we sit and eat, thinking of other things to talk about, like the football.

~

When we get home from David's party, I have to go and check on some sheep that were stranded the day before on a hill surrounded by floodwater. They were safe, but I need to see they are all OK. I am glad to be back on our old farm in the fells. The grassy banks by the beck are flattened, showing where the level of water had risen. Glassy puddles in the mown hayfields form mirrors reflecting the sky. Everything is clean and bright. The eggshell-blue sky is broken with white-cloud islands. As I stride downstream, the beck fattens from a trickle to something increasingly like a small river.

Our land is coming back to life again in countless little ways. Wildflowers hang over the fresh flood-bitten banks and rise from the gravel in its bed. Foxgloves tremble in the breeze. Little brown, orange, and white moths and butterflies flit to and fro across the clumps of flowers. A heron rises awkwardly from the rushes a hundred yards away, and twists and tumbles down the valley on its sail-like wings, gliding back to the beck two fields away. Our valley always looks shiny and new after a flood. But unlike in those biblical floods that swept away the sinful past, our waters have receded and the all-too-human jumble of the past and the present emerges once more.

~

Our land is like a poem, in a patchwork landscape of other poems, written by hundreds of people, both those here now and the many hundreds who came before us, with each generation adding new layers of meaning and experience. And the poem, if you can read it, tells a complex truth. It has both moments of great beauty and of heartbreak. It tells of human triumph and failings, of what is good in people and what is flawed; and what we need, and how in our greed we can destroy precious things. It tells of what stays the same, and what changes; and of honest, hardworking folk, clinging on over countless generations, to avoid being swept away by the giant waves of a storm as the world changes. It is also the story of those who lost their grip and were swept away from the land, but who still care, and are now trying to find their way home.

~

It is now five years since my father died, but we had left his rented farm long before. When we drove away for the last time, I shed quiet tears for the place where I had spent much of my childhood and youth. Those fields had been my playground, my classroom, and my first place of work. I left that day with a head full of memories of things learned and of countless moments of beauty. That farm was the prism through which I learned to see the world: the place where I was taught about the weather, farming, work, community, and the wild things around us. It felt like saying goodbye to a dear friend for the last time. I was abandoning a much-loved place to fall apart. But that farm was not our land and never would be. If we had a chance of holding on and doing anything good, then it would be on our own farm in the hills.

We made our home instead on my grandfather's farm with its 185 acres of owned land. Fifty years ago, it would have been a decent-sized farm, but by the standards of industrial agriculture today it is tiny. Our bank manager says it is too big for a hobby; too small to make much money. But it is ours.

~

I still travel back to the old rented farm sometimes. It is like visiting someone you once loved but who has now become a stranger. The gorse on the hillside that my father used to burn every few years has gone, scraped away with a large digger or bulldozer. The farmers there have brought that old field, the Quarry field, into the modern age. The hay meadows have been plowed out and "improved." There are no fields of barley, oats, or turnips. It is all tidier, bright green, reseeded, and farmed more efficiently. The linnets that used to flit upward to the telephone lines, as we played hide-and-seek; the rabbits that raced across the gaps between hawthorn clumps; and the yellowhammers that used to sing from the top of the yellow gorse bushes—they have all gone.

I remember my old man standing, wild-eyed, surrounded by flames as he set the crackling, tinder-dry gorse on fire. And I want to go down to the village to see the head teacher's wife who complained all those years ago and tell her the birds were better off with my dad and the old farming ways, and the burning every ten years. But she has gone too, and the world has changed, and it isn't my place anymore.

~

There is an old saying that we should farm as if we are going to live for a thousand years. The idea is that we might protect our natural resources better if we had to face the long-term consequences of our actions instead of passing on a mess for someone else to sort out. I find the thought of a thousand years in the future rather daunting and impossible to comprehend. Who is rich enough to be that holy? A few aristocrats or large conservation organizations, perhaps? No one with any sense doubts the need for change. The statistics on the decline of wildlife on British farmland in my lifetime are damning, and the climate is doing weird, terrifying, unprecedented things. We know that change is needed, but we have to work out what that change should be, and how we can deliver it—when we are neck-deep in the realities of this age.

We live on the earth; we cannot float above it like angels or separate ourselves from it entirely. Such misguided idealism emerged from the Enlightenment and intensified during the Industrial Revolution. People left the land for towns and cities, and then, a generation or two later, when they were better educated and more affluent, they returned to it as a form of leisure and escapism, developing a new kind of relationship with the landscape. When we left, we were farmers. When we returned, other people, tougher people, were the farmers, and we just loved "nature." We had become free of the harsh realities and were then several steps removed from what others now did in our name to feed us. We found the old ways hard to stomach and sought to escape from slaughter and death. Such utopianism speaks to our better selves, but there is a very thin line between idealism and bullshit.

Technically, the best thing we could do for nature in most landscapes is to not be there at all. Ecological despair at the scale of our impact has led some to suggest we should do just that, and embrace the worst intensive farming practices in some places in order to "spare" as much land as possible for wild nature in others (e.g., in the uplands). While such desperation is understandable, this potential solution won't actually work: the theoretical "spared land" rarely becomes the wilderness that is promised (the glorifying of "efficiency" involved actually ensures farmers everywhere embrace intensification using the huge arsenal of chemical and mechanized tools, completely sterilizing their fields in the process). Even if this weren't the case, large tracts of the countryside couldn't be wild in any meaningful sense because of essential human infrastructure: roads, railways, and housing. And, even if wild uplands were possible, these places would never exist separately as isolated ecosystems. To be anything remotely approximating "wild" there would require seasonal migration of large herbivores in large numbers between the uplands and lowlands, and large carnivores moving them around the country. That kind of truly wild landscape (and some of the species that made it function) has gone and isn't coming back. Abandoning farmland isn't remotely the same thing as restoring a wild ecosystem—plagues of deer replace hordes of sheep and little good is achieved. We might wish it was otherwise, but humans are the top predator, and either they can provide that function in an enlightened nature-mimicking way or species like deer or wild boar will wreak ecological havoc in the absence of predation.

Thankfully, we don't have to re-create an impossible past, as many species actually thrive in traditional farmland—especially in areas like hay meadows, coppiced woodland, and hedgerows. Many of our farmland species have now become rare, so we need to be careful not to lose the remnants of traditional farmed landscape that remain. A bird like a curlew has existed alongside mankind for so long that no one is quite sure where it lived in the distant wilder past. This isn't to say that we don't need wild places, and big ones—of course we do—but the complicated truth is that we need nature everywhere, even in our most intensive farmlands. To make them sustainable, we need to find workable compromises. In most of our landscapes there will never be perfect single-use solutions—pure wilderness or pure productivity. We need to put farming and nature back together, not drive them further apart.

England has a population density of over 1,100 people per square mile, and 56 million people to feed three times a day. Most of England is farmed, and realistically will continue to be, so our biggest and most vital ecological challenge is how to make productive farms much better places for nature. We can't wipe the slate clean, but every farmer can work up from where we are and make it much better. Some of the answers lie in the past—in how we farmed before we could cheat with new technologies. Other solutions require new ways of doing things, based on science (for example, by analyzing soil health, studying grazing practices to see what works best, and learning from ecologists about the habitats and natural processes we need to re-create). We can work the land and still have healthy soil, rivers, wetlands, woodland, and scrub. We can have fields full of wildflowers and grasses, swarming with insects, butterflies, and birds. We just have to want this enough to legislate for it and pay for it. To do this we have to opt out of the cheap food dogma that has driven farming and food policy for the past few decades. We might have to stop gullibly accepting every new technology and new ideology, and care instead about some fairly simple old technologies and ideas—like valuing good rotational mixed farming and enlightened land stewardship. One of the best ways to create a better rural landscape is to mobilize the farmers, and other country people, working with what is left of their old culture of stewardship, and tapping into their love and pride in their land. We can build a new English pastoral: not a utopia, but somewhere decent for us all.

~

In his final weeks, my father liked to travel around the farm with me and talk openly about what would happen after his last day had come. In this scenario, to my great discomfort, I had been promoted and he was no longer of any significance. He had become the past tense. I tried to stop him talking like that. The farm was still "his" in my mind, and I wanted to believe that he might not die. But he was clear that he had done his bit, carried the baton forward, and handed it on. The handing on, rather than being sad, or an ending, or defeat, seemed to give him great comfort, as he felt he had done what he had set out to do.

In the preceding months, when he found out he wasn't going to recover, and maybe was just a few weeks from death, he had to choose how to spend his remaining days. My mother asked him what he wanted to do, and the answer was simple (and entirely unsurprising). He wanted to go home to the farm to carry on his life for as long as he could. He made an unwritten, and unspoken, list of all the jobs that needed to be done on the farm. Then he worked, slowly, with failing strength at times, through his list.

One Saturday he took my elder son, Isaac, and my younger daughter, Bea, with him to hang a gate, because, he said, I would "never get it done." The gate was on the pastures opposite our farmhouse, between a field where our sheep were grazing and a meadow full of wildflowers and vetches that we mowed each summer for hay. It is flanked on both sides by an old wall covered in moss and lichen that shines, depending on the light and the hour, green, yellow, purple, and silver. Over the centuries the old wall has sunk a little under its own weight back into the earth, six inches down in some places. And it has sagged left and right, so it rises and falls, and meanders ever so slightly, like a loose shoelace. On the wind-wake side of the wall there lay a drift of last autumn's copper-bronze beech leaves, wind-brittle and crunchy like plastic crisp packets underfoot.

My seven-year-old son, Isaac, now calls it "Grandad's Gate." He tells me often that he, his sister, and his grandfather mended it. His grandfather spliced together two broken gates, recycling them, to make a new one. My wife had smashed one of those broken gates on the school run, reversing her car out of the farmyard without seeing it; it had blown half-shut as the children distracted her with their bickering in the back seats. The resulting mended gate looks a little peculiar, but it works perfectly. I pass through it almost every day. It swings good and true. It rained the day they mended it—but they didn't seem to mind. They all came back soaked but with smiles on their faces, proud that they had done it together. Dad knew there would be a million things he would not now get to teach my children, that so much that might have been would now never happen, but this day he had shown a little farm boy and girl that you could make something new from something broken.

~

It took me two or three years after my father's death to understand the simplest of truths: that I had just taken his place. We each have but a fleeting moment between those who came before and those who come afterward. I am now the one who makes the decisions that shape our farm, but my choices are almost as constrained as they were for the first farmers who settled here. The hilly nature of our land, and its northern latitude, height above sea level, soil, temperate climate (affected by the Gulf Stream), and our growing season dictate many of the ways we must farm. A Lakeland fell farm like ours is always going to be primarily a livestock grazing farm; it is part of the three-quarters or so of the British landscape that is unsuitable for growing crops. Yet we still have bills to pay, debts, and obligations to family and community. We need to produce something from our land to sell.

Our farm is still about the hard work of breeding and selling stock recited on our fields of grass. We must sustain our land and ourselves, until it is time to pass it on. I have no idea if any of my children will want to take on this farm, or indeed whether any of them will be able to afford to do so. Life is complicated, and it's hard to guess what it will hold for them. I merely hope that they will all learn as much as they can about farming during their childhood, including a respect and love for the natural world. This valley is a gift of a place to grow up. Children can roam half-wild across the fields and play in the woods and becks.

We will continue farming as if, someday, one or more of

our children will take it on, and will worry about whether they actually do in the years to come. For now, we get up each day and just do the best we can.

~

I lie in bed in the early hours unable to sleep, my mind racing like a revving engine. The farm fills my thoughts with decisions and choices to be made, crowding out anything else. I am a bundle of worries, failings, and debts. I sleep with the window open so that I can hear the sounds of the night—owls hissing, the wind in the ash trees, the geese talking to one another as they pass over. The light from the rising sun floods in all at once and lights up the silvering oak beams above me, casting triangular shadows on the plastered ceiling. The room is full of a glow like sunlight through a jar of golden honey.

I head downstairs. My son Isaac often follows, half-dressed, sleepy-eyed, and scruffy-haired, to go out to the fields with me at first light, but today he sleeps on. I must get out and see that a cow that is due to give birth is OK. She seemed to have "slipped" (opened) her pelvis last night, as if she was about to calve. I straighten up outside the front door, before a pale sky bordered with blue-green trees, and head out into the chill of the pre-dawn.

When I was about six years old I had an illustrated pop-up book about Puss in Boots. As the pages opened, the trees and mountains rose out of them to stand up in three or four uprights of cardboard scenery. As I look away to Patterdale from our farmhouse, the skyline reminds me of that book. The blue above promises a fine late-May morning. I walk up the hill to my barn, and see below me pockets of white mist lurking in the hollows: a milky ocean enclosed by the fells. On mornings such as these you can feel the warmth of the air on your cheeks as you climb the fellsides out of the basin of colder ground. The farm is visible again at first light, and I hold my breath a little for what will come. The mist below begins to burn away. The upper branches of the oak trees by our barn tremble in the dawn; the fell tops are glowing orange.

~

We have always gone shepherding at first light—by which I mean a tour of the farm to check that all is well with the cattle and sheep. That "stockmanship" remains a vital part of who we are, and is as necessary as it ever was. But my morning rounds of the animals have broadened now to studying our land and our valley, and trying to truly see and appreciate the nature in our fields and around them and how we can care for it more effectively. I try to understand the things that we once took for granted, or didn't see at all—the living soil, whether we are utilizing photosynthesis as well as we can with our grazing, the breadth and variety of wildflowers and grasses, and the woodland beside our fields.

We have not used artificial fertilizer on our farm for more than five years, and rely solely on healthy soil and sunlight. We are changing our farm in lots of little ways, creating new habitats by our becks, restoring bogs and grazing pastures in ways that will encourage a wider array of flowers, restoring meadows by bringing back seeds and plants of species we lost in the past century, and planting more woodland and hedgerows. I have come to care about half-invisible but vital things that we never thought about in my childhood, like moths, worms, dung beetles, bats, flies, and the wriggling life beneath the rocks in our streams.

Two fields away the curlews rise to the warming air and on the fell a cuckoo calls gently in the shadows of the woods. Cuckoo, cuckoo, cuckoo. Rooks caw angrily at one another from the sycamore trees by the houses down at Bald Howe. And jackdaws clack and chatter somewhere in the waking dawn. Twenty strides down the field from the barn, with my dog Tan by my side, I am soon back beneath the coming day, in a cold field, in the shadows again. It will take another hour for the sunshine to angle down into the blue-gray valley bottoms. Tan is wet, speckled with white grass seeds and yellow petals. The cold claws at my unbuttoned neck, makes my skin shiver; my leather boots are sodden.

The sunlight is almost cresting the brow of the fell to the southeast, a promised line of orange, lemon, and white. But here beneath the new day the delicate flowers are still closed tight, shut like little fists clenched from the cold. Buttercups bend their heads down in submission to the night; dandelions are bedraggled with dew. A roe deer doe with a fawn hidden somewhere in the meadow rises and gallops off to the field boundary as we pass, wary, but held by magnetism to her young. In the next field the cow lies chewing her cud, with her new calf tucked alongside her. Her shiny teats have been suckled. The cow smiles at me, as if she realizes I have been worrying for nothing. When I step closer, she snorts a little and shakes her head, to let me know that my interference is neither necessary nor useful. I back off.

~

I have worked here my whole life, but I am only now beginning to truly know this piece of land. I stumble across a field at a different time of day, or in different light, and I feel as if I have never seen it before—not the way it is now. The more I learn about it, the more beautiful our farm and valley become. It pains me to ever be away; I never want to be wrenched from this place and its constant motion. The longer I am here, the clearer I hear the music of this valley: the Jenny wren in the undergrowth; the Scots pines creaking and groaning in the wind; the meadow grasses whispering. The distinction between me and this place blurs until I become part of it, and when they set me in the earth here, it will be the conclusion of a lifelong story of return. The "I" and the "me" fade away, erode with each passing day, until it is an effort to remember who I am and why I am supposed to matter. The modern world worships the idea of the self, the individual, but it is a gilded cage: there is another kind of freedom in becoming absorbed in a little life on the land. In a noisy age, I think perhaps trying to live quietly might be a virtue.

~

A pair of ravens honk and rattle through the half-lit sky. Each beat of wings rasps the air and sounds like an old man wheezing for breath. They see me as I climb over the wall back into the meadow, and give their alarm call and are soon gone. This field doesn't look anything special in winter, not noticeably different from more intensively farmed land a few miles away in the lowlands. But three weeks ago I closed it off from grazing mouths to grow our hay, and it has become a place of amazing ragged beauty. Each day it looks different, passing through several colors and stages as over a hundred different plants race one another to the sky, flower, and then seed.

I climb the fence at the meadow's edge to take the "scenic route" home along the beck. I disturb two mallard ducks from a sandbank where the sheep gather to drink and they explode into the sky. Behind me, the fresh-calved cow moos to her friends in the herd, and they reply.

I would once have thought of these fenced-off, largely ungrazed strips of land as wasted, but they have become my classroom for understanding how wilder vegetation develops over time, and how rivers change when they are allowed to, and how trees grow, age, die, and rot and are recycled into the earth. There was nowhere like this on our farm a decade ago. They exist because of a young woman called Lucy, who changed my father's and my views of what it means to be a good steward of this land.

~

Lucy came to see us for a meeting about our rivers. Dad said she was from the "Water Board." She wasn't. Everyone associated with rivers or water was described as being from the "Water Board" and held in mild disdain. She was actually from a local river conservation charity. We met in the caravan that passed as our makeshift farmhouse before we had converted my grandfather's barn to live in. My dad rather enjoyed having this caravan by the roadside; he called it *his* caravan despite us paying for it. He would invite friends in for a coffee or a can of beer and a chat—they called it the "Matterdale Social Club."

At that time Helen and I were living in Carlisle in the only housing we could afford, a redbrick terraced house among what had once been cotton mills and factories. I drove back to the farm to work each day. That year in the city there were devastating floods, and the waters reached to within a few meters of our front door. We saw hundreds of people displaced by the muddy water. And when the water receded, their soaked and smelly possessions were ripped out and heaped in the streets. Wide-screen TVs.

Carpets. Chairs. DVDs. Tables and chairs. Rugs. Children's toys. Things being thrown out of doorways and windows and into dumpsters. There was talk about where this water had come from, and what might be done about it. Our farm was upstream, twenty or thirty miles away, a tiny part of a huge watershed—and I think that's why Lucy came, as part of the process of looking at the landscape and exploring whether some of that water could be slowed down to reduce the floods.

We sat inside the caravan with the gas fire on, waiting for her to come and talk to us about our rivers. My dad was joking around as if she was coming to tell us off and he wasn't going to take it too seriously. Or else she would jail us for some crime we didn't yet know we'd committed.

She came in and braved a mug of tea. My father wasn't known for his kitchen hygiene—the mugs were filthy. He often swilled the tea-stained cups out with some suspect-looking water. Lucy was disarmingly normal and likable, quite unfazed. She persuaded us that she wasn't here to catch us out or tell us off. After we had exchanged pleasantries, the rain paused and we were able to walk our land. She talked to us about how unnatural most rivers were, and what needed to be done about it. She showed us some good and not-so-good things about our streams, without sounding lofty or know-it-all. She told us that the becks that we had always known, and rarely given a thought to, had actually been straightened and dredged, often by hand, in the nineteenth century. They were really man-made drainage channels—too straight, too deep, and too regulated to be much use for salmon and trout. A healthy stream, she explained, needs slow bits and fast bits, wide bits and narrow bits, shallow riffs and deep pools, and places where it drops gravel and silt to form places where fish can spawn. There was no judgment or condemnation in this. She just told us it straight. My father and I soaked it up, because no one had ever bothered to explain all this to us before. It made sense.

Lucy said there were lots of sensible things we could do to "slow the flow" of water from our farm downstream. Rivers needed to be made more natural once again, less engineered, and to be given space to wiggle and meander, with banks covered in trees, with lots of old wood in them to help the rivers become healthier for fish and other creatures. Planting trees in the right place, she said, could also slow down floodwater, and let it soak into the ground instead of running off the surface. And fells and moors "roughed up" with tussocks and healthy mosses and peat bogs could act like giant sponges. Healthy soil in the pasture fields could hold water better than compacted, dead soil. She said that if these things came together to create more natural river habitats, then it would make a big difference to our whole landscape. It was perhaps unrealistic to think we could do all of these things everywhere—but there were many places where they could be done, including on land like ours. She asked lots of sensible questions about how we farmed and talked us through what we might be able to do on our land differently if they helped us. And it wasn't a fantasy. She had funds. She could pay for things *if* we would work with her.

Like many farmers we were perhaps still in denial about how much our land had changed. Even on Grandad's beloved old fell farm things had altered over time—not as radically as on the more improved land beyond the valley, admittedly, but the same forces were gnawing away at it all. Between our little fields the old hedges had grown out, and the fences were rusty and the posts rotten. We couldn't afford to maintain or replace them, so our small fields had been amalgamated into bigger ones, all managed the same way and often bare-grazed with sheep.

Lucy offered to help pay for us to put all those old field boundaries back, and many new ones, so our farm would become a quilt of smaller fields once more. But she wanted the riverbanks, boggy bits, and woodland fenced off and managed differently, with only light and occasional grazing. She wanted wider hedgerows and thousands of trees planted.

We had been with Lucy for about an hour when it occurred to me that something odd was happening. My family didn't really have these kinds of conversations. We were always suspicious of meddling outsiders. But for some reason my dad was quite enthusiastic. The atmosphere was constructive and respectful. Lucy's vision of a valley that was better for rivers and fish had a lot of overlap with our thoughts about how much better the old farming landscape had been. Her plan would help us to reconstruct the original field pattern around our new family home and restore the fell farm that had begun to slip away. If we gave her what she wanted, we would be getting a load of new fencing that would be half paid for. There was a chunk of selfishness in our listening. We were probably more interested in the half-price fences than what might happen inside them.

My father heard all this, and then, to my surprise, pushed me to make the final decision, saying, "It will be his farm. He can decide." When I asked him later for his approval for changing our farm quite radically—becks fenced off, trees planted, a mile of new hedges, and fields restructured—he said we should just "get on with it, unless you've got a gold mine to pay for all the new fences yourself." Weeks later a team of Polish men came in the middle of winter and began to alter the farm to the plan agreed by us with Lucy. The Poles worked through the snow, impressing my father greatly with their hardiness.

~

With hindsight, I can see that Lucy did something very clever indeed when she persuaded us to fence off those becks. And perhaps my father can have a little credit too for giving those changes his quiet blessing. Lucy didn't take those bits of land away or end our responsibility for them—she just put them in our care with a different set of rules. We would, for the first time, manage some land for something other than farming. These strips of land would be almost entirely given back to nature to do its thing. Though I didn't realize it then, she and my father made me the guardian of some half-wild spaces. This was a big shift for a farmer, not an easy one initially, but revolutionary in the longer run. It was probably the first time in generations we had "unimproved" land. This first step changed me as well as our farm. Once you start moving, the steps get easier.

~

Soon after the Polish workers had fenced off our riverbanks, the grass grew rank and much rougher than it had ever done before, and there was an explosion of voles. They scurried between clumps of grass, from tree roots to rocks. Within weeks, barn owls returned to our land after many years of absence, to feast on the voles. Everyone in our family felt proud about the owls, as if we had been paid back straightaway. It felt true to who we were, something my grandfather might have appreciated.

After that, wilder plant life began to take hold, though it took longer to emerge and was harder to see. For the first three or four years, strong, coarse grasses shot up on the banks and choked out the smaller delicate flowering plants; for a while it became less biodiverse. But in the fourth and fifth years purple and yellow flowers were suddenly blooming everywhere, rising above the grasses, and a golden haze of insects, butterflies, and bees flitted to and fro over the wilder areas as the summer sun set. The little becks began slowly to find their own space again, with a little help from us, breaking out of their nineteenth- and twentieth-century straitjacket channels, and twisting and turning like spaghetti thrown on the floor. Little willow and alder self-seeded on the banks, and behind them little gravel beds began to form, tiny at first, but soon fattening until they were changing the course of the beck. As the water slowed behind the saplings, it dropped finer gravel on which fish spawned.

~

The riverbanks are now bustling little highways for wildlife. They are rich in purple and yellow and pink flowers, moths, butterflies, and stoats, and crossed with the sunken paths of hares, badgers, and foxes. An otter now claims part of this promising territory with her two kits. All around me is birdsong—chaffinches, blackbirds, thrushes, warblers, blue tits, and coal tits. Wood pigeons coo. As I near the house for my breakfast, the roe deer with her fawn dashes away and up past our farmhouse: this is her route from the meadows to the woodland beyond.

~

My grandfather told me to "plant trees" when I was ten years old. It was a slightly baffling instruction because to the best of my knowledge he had never planted one. I think it was one of those things he knew he should have done. He spoke with great affection about the Scots pines that once grew above his stackyard when he took the rented farm on in the 1940s. But I only knew their rotting stumps—where I had played with plastic toy soldiers until I was bitten by the red ants that spewed out of the decaying wood when I broke off a piece by mistake. But over the years those words echoed in my head and did their work.

~

Last March, on a mild gray Saturday before lambing time, we gathered up the children and went out to a fenced-off wilder area on our land. The little oak saplings peeked out of the burlap sacks and were starting to bud. We had to get them in the ground. Our spades slipped into the turf and levered open crevices for the roots. Gently, we posted the eighteen-inch-high saplings into the open earth, then pressed the turf and soil firm around them with the heels and toes of our boots. My elder daughter, Molly, carried the tree guards and stakes. She secured each sapling with a guard to stop them from being munched by deer. She dawdled as she passed the beck, watching some little fish dart for the shadows. Isaac stumbled across the tussocks of long grass with a handful of saplings. He told me he would come back here when he was a grandfather to see if these trees "have actually grown, or not." He sounded doubtful about the whole enterprise.

My mother was away across the beck, untying the bundles of hazel and hawthorn saplings. Her hair was grayer now, and she wasn't as strong, but she loved to help us on the farm with lighter work when she could. She said she felt closer to my father when she was here, that he was somehow still working among us. His ashes were scattered high on the field called the Horse Pasture, so that he could "keep an eye on us." When she was grieving and broken, being on the farm doing the work he once did seemed to help. She became the memory of this farm and reminded me that my father always left a straight-trunked little oak or ash tree every thirty or forty feet in any hedge he was laying, so it could grow up beyond the rest of the hedge. And in the past year when we laid some of these hedges again the traditional way, after thirty years of neglect, the half-grown trees left standing were those my father had let grow the last time. They stand proudly above the hedge, their crowns billowing out to touch leaves with their neighbors along the dike.

~

At the top of the little woodland corridor, I pass by the "cloven stone," a giant boulder deposited here thousands of years ago by a glacier, and through the wooden gate to our house. The valley is now carved in half by the sun. Behind the farmstead the south-facing fellside shimmers in the new day. A hare leaps through the meadow beyond the beck. It rises and falls, like a carousel horse seen above the crowd at a fair, its top half appearing and disappearing with each grass-bound leap. And then it stops to watch me pass, ears flicked backward like a capital K poking out from the grass.

~

Back at the farmhouse a small boy is peering out of the window, a sturdy little man with a mop of blond hair and a wicked, hopeful smile. Our youngest, Tom, is not yet two years old. He seems to be as stubborn as the grandfather he was named after, and as obsessed with the farm. He presses his face up to the glass, fascinated by what lies outside. As a baby of a few months old, he would sit watching the hens in the garden, or the red squirrels stealing the birds' nuts, or shout "baa" loudly at the sheep on the hillside opposite. He is furious if I don't take him each day to feed the sheep and cattle. Left behind, he stands at the door, tears flowing down his cheeks, or shakes the garden gate in his fury, howling at the injustice of it all. Something beyond the house and garden calls to him. Often he does come with me, and his mother sometimes comes along as well to keep him safe while I work. He seems oblivious to bad weather in his jumpsuit and mittens, and some days sits on the quad bike, overalls soaking wet, his little cheeks cold and red, but set with a steely determination to be there, come what may.

When I shout the flock up for hay, he shouts too. When I shout commands to the dogs, he copies at the top of his voice—"LIE DOWN, TAN." The dogs lick his face despite his attempts to swipe them away. After we have been around the flocks of sheep and fed our cattle, we head home. When I hold out my hands to lift him from the quad bike, he shakes his head, as if he would happily sit and wait in the rain all night rather than miss it tomorrow. So as I enter the farmhouse, I sweep him up in my arms and tell him I am sorry, and he can come and see the new calf later. I have to give a full report of all I have seen while I eat my breakfast. But he seems skeptical about the dragons in my tale.

~

After breakfast I get Tan and my other dog, Floss, and we head up the wooded gill by the farmhouse. This giant V-shaped ravine, carved out of the hillside by the beck, connects our lower ground to our highest fields. In storms, the tumbling stream sings down past our house. In winter, wind roars off the fells and over our roof, wave after wave, over the slates, with us hunkered beneath it like barnacles clinging to the rocks. The gill is almost a quarter mile long and perhaps seventy feet deep, at its deepest point. Spruce trees, planted by some unknown hand a century or more ago, have grown from the bottom of the ravine and fill it with their evergreen branches. From the high ground you can look into the crowns of these great trees. Buzzards nest in one, on a platform the size of a child's bed, made of branches. When these trees fall or blow over in a storm, they crash down to the other bank of the gill and create mossy aerial walkways that the red squirrels tread smooth with their feet. I duck under them as I make my way up with the dogs. These gills are vital ribbons stretching through our farm from the wilder woodland above our land to the valley floor. The twisted and broken old trees seem to own this place. Isaac says that perhaps they come alive when we aren't looking, like the Ents in the Tolkien books. Their tangled roots sometimes reach out of the ground and trip me up, scattering mice and their enemies the stoats. As I reach the top and emerge into the sunshine, I can hear a post hammer striking somewhere—a wooden crack that echoes around the fells.

~

I pass through the gate into the New Field, our highest field. It is poor ground, two-thirds a steep, south-facing grassy bank, and one-third of bog. We bought this in the months after my father's death, because it came up for sale. It was cheap to buy, and with good reason. It is only suitable for rough cattle and sheep grazing, and not much use to anyone except us because of its isolated location up the lane above our farm. The banks of it, in summer, are a mass of flowering grasses, pignut, red and white clovers, bugle, buttercups, forget-me-not, cowslips, vetches, orchids, and burr thistles. The field was bought with set environmental scheme rules about when it can be grazed and how, in order to protect these wildflowers. The bog at the bottom is so sodden that you can scarcely stand on it; the turf suspended, but only just, like thick skin on custard. The higher land all around it drains down into this basin, and because the water mostly cannot escape, it is always wet, a peaty mass of rushes and long grasses. Little fish, bullheads, dart away from my shadow and dive down into the pondweed a few yards away.

The fences around this field were mostly redone before we bought it, but one, across the bog, was falling down and had been left, perhaps because the fencing contractor didn't fancy getting his feet wet. So, when it became ours, I hired a different gang to come and build a new fence to make the land secure. When we walked the fence line to measure the job and agree on a price, we found a dead ewe of my neighbor's lying stuck in the bog, face chewed off by badgers. I half expected the lads to tell me to do the fence myself, but they welcomed the challenge and now they are up there working. I have come to see if they are OK.

Half a dozen sheepdogs are playing in the sunshine on the grassy banks, chasing each other and fighting. A broad-shouldered, black-haired lad wades through the bog toward me. His bare, suntanned arms are wet with sweat, as he lugs fence posts on his shoulders across the rushes to where his mates are working. With his free hand he swipes at a fly that has nipped his arm. These damp places with long grass in them are hellish for horseflies, or "cleggs," as we call them. They suck your blood and leave a giant welt on your skin. Others are hovering gently around him for a chance to land. I wade out to meet him at the line of newly sunk-in fence posts, sodden to my knees, and as I reach him I see that the biggest of the lads using the post hammer is standing up to his belly in dark water. They are using very long posts to grab into harder earth down below, and the tension of the wire to hold the fence in place. The black-haired lad tells me they have seen some stray sheep of ours on the "lonnin," or lane: two ewes and three lambs. And some of the sheep on the enclosed land beside them are being bothered with flies and will need attention from me.

When these lads are not doing contract work, they gather the fells for other farmers with their sheepdogs, bringing down the flocks. They also shear sheep and work on their families' farms—the kind of people who made this landscape. They look heavy-legged, but are having fun working together. The lads tease one another about their girlfriends, or failed attempts at getting one, and about the biggest of them, who is soaked and demanding they go for a KFC in the local town at the end of the day. He loves that shit, his mate tells me. Then rain falls, silver and heavy, catching in the brilliant sunlight like millions of droplets of mercury. We shelter under a willow tree. The rain thwacks on the olive-green leaves. We wait.

Blue and emerald dragonflies dart about under the outstretched limbs of the ash trees. Little ripples swell outward from the fat raindrops as they splash down on the bog. A newt hangs listlessly a few inches beneath the surface of the water, ignoring the rain, with arms and fingers outstretched, watching us. The shower passes. The lads wade away for more posts. Look, says one of them to me, and points. Isaac comes running across the wet field, his legs brushing the flowers as he runs. He passes through the sheep lying in the grass, chewing their cuds. I shout for him to wait for me on the dry bank, and head out to meet him. Floss gets there first and gets a cuddle. We four head after the stray sheep along the lane that rises and falls, and twists and turns, from our land to the distant fells. Strays are a pain in the backside, but if they haven't gone far it is better to retrieve them now before they cause more trouble in the gardens of the village.

~

Once we crest the first hillock we can see the escaped ewes and lambs threading their way along the lane. I send Tan to pass them. He's off like a bullet down the ruts, until he is on them, and past their flanks. They turn back and run to us. Isaac opens the gate for them into our field called the Top Rigg and Floss turns them in. The sheep gallop off down the field. They flush a brown hare from its form, and it bolts. Hares love this open grassland. In spring they chase one another in mad day-long races across these fields, and then stand on their back feet to box. They give birth in the fields and tuck their leverets into the rushes and tussocks of grass. They sit motionless, with big hazel-colored eyes, watching us pass.

Up here the land is more open—almost moorland—with only an occasional hawthorn dike to break the skyline. These fields are pastures that have been managed the same way for countless decades. They are unsuitable for growing crops. Our ewes with two lambs graze these fields in summer, while those with one strong lamb go back to the fells where they can be reared on the tough mountain vegetation. Some cattle have gathered behind us, reaching their heads over the gate from my neighbor's land. They belong to my friend Alan. He is one of the elders of our community, a man of good sense, from the same forge as my father. We often swap notes over the fence about the price of sheep, the weather, or the wild things we have seen. He seems sad at how things have changed and points out that we receive now (in real terms) a quarter of what he was paid as a young man for his sheep. The value has been stripped out of good farming, by cheapening food too much.

And yet our animals are still what keep us here, creating the work and the income that helps pay our bills. We generate two or three times as much income from sheep and cattle as we do from environmental payments (despite being in the highest possible levels of the available schemes). Our lamb and beef do not earn the premium that would compensate us properly for producing them in ways that are less efficient but good for wild plants, insects, birds, and other animals. I see no prospect of this in an age when everything revolves around producing cheap food. Meat has been reduced to a pitifully low-priced commodity in supermarkets, when it ought to be a thing that is respected and valued—even if that means eating less of it. Many British people have lost the taste (and cooking skills) for things that can be produced sustainably in their own landscapes. The mutton from the older sheep that have reached the end of their breeding lives on the fells would not have a market at all if it weren't for the British Asian community, who, thankfully, still have the taste for cooking and eating it. The old farmers in our valleys know exactly when the festival of Eid starts and ends because they time their sheep-selling to coincide with the feasting.

~

When I was young and first began to realize how fucked up and complicated the world was, I thought we could turn our backs on it all and hide away on our farm as if it was a kind of cocoon. And when I reluctantly left to go to university, and to earn my living in that wider world in my twenties and thirties—mostly sat in front of a computer doing what other people told me to—I loved having a place

I could retreat to, to ground myself in work that to me felt real. The farm kept me sane.

When people dream of getting "back to the land," they often have this kind of escapism in mind. If you have been parked in a cubicle in an office for years of your life, with lengthy commutes on trains or in traffic jams, then perhaps a farming life looks like freedom. But I've come to see that the reality of being a farmer is anything but an escape from the world; it is often like being a slave to it. Everything that happens on a farm is affected by the era it exists in; it is shaped by a host of powerful external forces. We are dangled like puppets, pulled to and fro by invisible threads. Somewhere, just out of sight, those threads are connected to how you shop, eat, and vote—as they always have been. But in the past fifty years we have let those strings be pulled by supermarkets and other large corporations until most farmers have been reduced to low-price commodity producers with very little bargaining power. We are struggling to face up to the ecological disaster this has created, at the same time as producing the cheapest food in history. And instead of addressing the structural problems of a food system in which almost all the power and the profit is taken by large corporations that care little for the health of citizens, farms, or ecosystems, politicians offer inadequate, thinly spread subsidies, or "environmental payments," to patch over its worst effects so that it can carry on.

~

My farming friends can now be crudely divided into three groups. A third of them have begun to change their farming, finding niches to earn money, and are committed to trying to be good ecological stewards. Another third are open to change but have limited room to maneuver, caught as they are in the financial realities of trying to run a profitable business, often as tenants and with heavy debt. The final third are deeply skeptical—or are still true believers in the intensive postwar model of farming. Talk is cheap, they say. They are focused instead on delivering what society pays for, in supermarkets—cheap commodity foodstuffs.

Our leading agricultural colleges still churn out "business-focused" young farmers, fired up with productive zeal. Students are taught to be at the cutting edge of the new farming, applying science and technology to control nature. They are taught to think about the land like economists. They are taught nothing about tradition, community, or ecological limits. Rachel Carson isn't on the curriculum. Different colleges and courses elsewhere churn out young ecologists who know nothing about farming or rural lives. Education is divided by specialism, and sorts the young people into two separate tribes who can barely understand each other.

An agricultural student came to our farm last year, and when I showed him our hay meadows and explained their diversity of wildflowers and grasses, he looked at me with a mixture of confusion and contempt, as if I were a charming but delusional fool from another age. He boldly told me his tutors would recommend my fields were plowed to get rid of the "weeds" and reseeded with more modern grass. The idea that a farm might be anything more than a productive business was completely alien to him.

Thankfully, my father never sent me to agricultural college. He was old-school and thought those places turned out people who knew the cost of everything and the value of nothing. In my early twenties, I remember telling him, admiringly, about the farm of a friend of mine that was doing lots of cutting-edge technological things, and he simply said, "Let's give them twenty-five years and see how they get on, before we get too carried away." Time was his test, not short-term profit or what was fashionable.

Agricultural education is still overwhelmingly about change and innovation, and "disruption," not what is sustainable and what will work in the long run. From the modernizing perspective, the student in my hay meadow was right. The current economics of farming are such that almost no genuinely sustainable farming is profitable at present. Farming for nature is economic suicide. Produce meat at a greater cost than intensively produced chicken or pork and you are considered an anachronism on the supermarket shelf.

I have to ignore my accounts in this bid for good husbandry and hope the rest of the world comes to its senses someday soon. Of course this is no basis for a sound system, but I decided years ago that if I had to work off the farm to top up our income, to enable me to look after our land properly, then I would. There is nothing new in having to adapt and earn a crust away from the farm. I know that if we are too proud, too stubborn, and too unbending, then we will be finished. We will have to learn some new tricks. But I won't ruin our farm trying to copy the conventional model of industrialized farming that I have come to see as destructive. In this I am as stubborn as my father.

~

Isaac and I walk down the lane and find the hole in the dike where the sheep escaped. There are telltale strands of wool on the thorns. Good fences, hedges, and walls help us to contain the sheep where we want them to be, but naturally the sheep have their own ideas. I shove some branches through the hedge, and some more through the other side, until it is crisscrossed with sticks. Isaac isn't sure it will turn them, so I shove in another branch and pull it until the whole patch is a rigid mass of thorns.

Our ewes and lambs are in three separate flocks, because that fits with the field pattern of our farm. We rotate them through the fields, allowing the grass to recover between grazing. Isaac, Tan, Floss, and I cross these fields. Swallows swoop past us just above the grass for the insects that our feet disturb. Isaac says they are like the X-wing Starfighters in the *Star Wars* movies, as he takes extra galloping steps to keep up with my stride. The ewes and lambs raise their heads when they see us, or trot off a few feet, but mostly they are not disturbed because they know us. There is no substitute for walking the land, for feeling it give gently (as healthy soil should) beneath my boots.

I try to explain to Isaac that we must decide how and when, and with which animals, we graze our fields: whether we graze in rotation around the paddocks, or permanently graze through "set stocking." How heavily we graze, and how much vegetation we leave behind, and how long we let the field recover for: long enough to grow three inches of lush grass as conventional farmers would do, or longer still for orchids and wildflowers to flower and seed over thirty or forty or more days. We are now trying to graze to increase biodiversity and create healthier soil, and that is different from our old ways. It requires longer periods of rest between grazing, and a greater mix of animals doing the grazing. The sheep don't always appreciate the longer older grass, liking the shorter sweeter stuff, so we need good fences and hedges to stop them willfully moving themselves, like the escapees on the lane.

Choices like these matter. They determine whether this land produces healthy living soil or whether it is eroding, compacting, and dying; whether we have wildflowers, insects, birds, and trees on our land, and how many; whether the hedges are bushy or losing their tangled hearts; whether the becks and wetland are wiggly or straight. When accumulated across whole landscapes, these choices determine what kind of countryside we get, and whether there is room for nature, or even people, within it. These very specific choices are rarely spoken of, shared, or understood outside the closed world of farming.

~

A buzzard circles above us in the blue, mewing to its mate. Isaac watches it ride the thermals to the hillside in the distance above our young cattle. They have their heads down grazing, and I can count seven of them and see all is well, so we don't need to go closer. Five years ago, we started to keep cattle again, close to twenty years after my parents' cattle were shot by a police marksman with a rifle during the foot-and-mouth outbreak. There is a lot of joy in building a new herd, in finding and buying the right foundation females, getting to know them, and seeing their calves being born and growing up. My father would have loved this, and I miss having his opinion, as he knew cattle better than I do.

We are building a herd of Belted Galloways—curly-haired black cows with big white belts around their bellies, from just across the Solway Firth estuary in Scotland. They can be seen from over a mile away with their distinctive broad white stripes. Cattle are a vital part of the farming patchwork that once existed throughout our valley; it had slowly been lost as we specialized and simplified our farming over the past thirty years. But I want my children to grow up with cows around. I want them to know about cows' different grazing habits, which are necessary for some of our new wood pastures and the riverbanks.

~

We traveled to meet a respected breeder and walk around her cows. Tom was on my shoulders. She told the children the name and story of each cow, and she scratched the cows' backs as they gathered curiously around us. A large bull wandered about among the cows, bellowing and kicking up dust like a wild bison on a prairie, and sniffing at their tails. Tom's little fist tightened on a handful of my hair. But the bull was harmless. He had been the champion at the elite sale at Castle Douglas the autumn before.

There was a strong sense in the field that the breeder would rather not sell any of them, but of course a farmer must. She offered us the choice between two beautiful in-calf heifers, and then quoted us an eye-watering price. It was clear the cows weren't going anywhere unless the price was right. We went home without either of them, slightly demoralized and realizing that our herd-building would take a lot longer and cost a lot more than we had envisaged. But you can't start a herd or a flock with poor females. So, three weeks later we rang and bought the one we liked and went to collect her. The following January she had our first pedigree calf named Racy Ghyll Lily the 1st.

~

I had forgotten how much I love cattle—their clannish ways, their friendliness, and their constant munching. I had forgotten how much a part of this place they are. Our Belted Galloways are properly at home in this northern landscape. They are square, chunky, and hairy, with short broad heads and rounded chests and bellies (you need a bit around the middle to digest plainer fare). In their two-layered winter coats, with heads covered in curls of hair, they can endure our cold and wet winters. They have been designed by a long history of selective breeding and evolutionary pressure to be hard as nails. They require no fancy buildings or equipment, and little more than a handful of hay on the worst of days. They grow leaner through winter but hold on to their flesh better than modern breeds that "melt" quickly when they face hardship; they know how to find enough to eat in the toughest of times.

It has been a revelation to me how they shape the landscape—clearing rushes and helping plants like orchids to spread with their hoofprints. They have patchier grazing habits than our sheep, meaning that they leave some areas of the field at key moments of the spring and summer to get longer, allowing some plants to flower and seed, and graze other places shorter, benefiting other plants that like light. Ground-nesting birds like curlews seem to like this mix of habitats, and are often to be found nesting in the cattle-grazed land. Without the cows, the thuggish rank grasses would outdo the smaller, more delicate, and rarer plants. The field where the cows graze all summer echoes with grasshopper song and grasshopper warblers, and is alive with clouds of moths and butterflies. Orchids rise in the grass in July. Mayflies dance above the pools where the cattle drink. Swallows hawk around, catching the flies that swarm the cattle muck. Rooks dig among it for worms, larvae, beetles, and grubs, and spread the nutrients across the field.

Our permanent sheep pastures, like the one we stand in now, have a different mix of almost fifty species of grasses and wildflowers that can endure more uniform grazing, and which thrive on a shorter sward. Birds like skylarks and meadow pipits love these shorter sheep-grazed fields. Two or three times a year I see a tiny hawk, a hobby, hunting them. So, the lesson is not so much that cattle are good, and sheep are bad, as that we need a healthy mosaic of diverse farmland habitats: our animals are the tools that can shape places for better or worse. And all of our grazed habitats are much more biodiverse than any intensive monocultural cropped farmland.

Isaac and I head home. I am soaked to the knees, Isaac to his middle. We kick through the molehills that litter the pasture as we go, spreading the soil across the grass.

~

Astonishingly, farmers over the centuries have not really known much about the biology of plants and soil. Soil was just the dirt we grew crops in, the stuff beneath the pasture surface, of no great significance in its own right. We took it for granted. We knew it had a pH score, but not that it was a living ecosystem. If soil was thought about at all, it was only to wonder whether we had knocked it into a fine enough seedbed with the harrow or whether it needed a top dressing of lime or artificial fertilizer. But if you had asked my father or grandfather, or me, what soil actually was, or how its "microbial life" worked, you would have been met with blank looks.

Isaac and I have learned about soil together over the past couple of years from our friend who teaches "regenerative agriculture." She taught us some simple things in our fields. She dug, with a spade, six-inch squares of soil and we counted how many worms there were in each. This showed us how much healthier the soil was in some places than others, and we discussed how our past farming had created those differences. Our teacher pushed a plastic ring into the ground and poured water into it to show us how much more water healthy soil holds, and soaks up quickly, than poor soil. And she forgave my ignorance about photosynthesis and patiently explained how much better plants are at growing when they have more leaf mass and deeper roots, than when they are always grazed short.

She has helped us understand that the living soil is the most important habitat on the farm—the basis of the food chain—and that all farming is ultimately "livestock farming," because whether the farming on the surface is a plant crop or an animal grazing we are always exploiting and utilizing incredible numbers of living things above and below the surface. About half the living things on earth live in the soil. It is its own world, with all kinds of strange and wonderful relationships between plant roots and algae, bacteria, nematodes, weevils, protozoa, fungi, and a host of other things I don't pretend to understand. A handful of healthy soil can contain more bacteria (and countless other tiny living things) than there are humans on earth. The good news for the soil on our farm is that, unlike on plowed and "improved" land, it is permanently covered in a thick and diverse carpet of grasses and wildflowers—meaning it is never completely exposed to the sun, rain, and wind, which would heat it up or blow it away or erode it. In some places, plant roots can reach down three feet deep, holding the soil together and providing pathways for nutrients.

The secret to keeping it healthy, we now know, is to mimic wild herbivore behavior with sudden bursts of mobbed grazing, which tramples the grass and dots it with shit, piss, and saliva. It looks trashed, with wasted grass trampled down, but it is soil heaven. Dung beetles, worms, and countless other creatures start to take the leaves and the herbivore muck (which is also full of condensed and partly digested plant matter) back into the ground. A giant feast ensues, a soil-based feeding frenzy, which we can encourage further by having more trees and hedges that litter the ground with leaf matter and rotting wood. Together they all play a valuable role in soil formation, and over time new soil gets created and carbon gets trapped in the ground.

Our soil is now safely held in place, with no erosion, and its subterranean ecosystem can thrive, free from fertilizers and free from the plow. The birds told me this was working, because they spent all their time feeding on the two fields I managed this way as an experiment four years ago. The soil is the base layer of our ecosystem.

Unfortunately, we can't all be fed from pastoral systems. The plow, and the annual crops it makes possible—corn, wheat, barley, soy, sorghum, cassava, potatoes, and rice—provide food for most human beings. But in the past thirty years we have learned that plowing is ecologically disastrous: it breaks open the soil's microbial networks, raises or lowers its temperature (by baring it of vegetation on its surface and exposing it to the elements), kills its bacteria and micro-organisms, and leads to wind and rain erosion on a massive and unsustainable scale. This news is staggering—and hard to take in for many farmers. Add into this mix the fact that artificial fertilizers and pesticides destroy much of the life in the soil, and farmers (and human beings in general) have a major problem on their hands. Our civilization rests on the plow (and the chemical tools of the postwar period) and yet the plow is a problem. So we are having to figure out how to farm in highly productive ways that work with, and respect, nature, and that means changing how we farm and thinking again about the tools we have become reliant on.

We can grow annual plant crops without the use of the

plow, drilling seeds directly into the ground with minimal soil disturbance (called "no-till farming"). But this brings new challenges, because how do you kill the harvested crop to make way for the next crop without using a plow to bury it, or chemical sprays to kill the regrowth or weeds? We might someday develop and grow perennial grain plant crops so that plowing becomes unnecessary (smart people are working on this, but it isn't a mainstream reality yet). And how do you fertilize the fields without using artificial fertilizers?

The solution to these challenges, in many places, is a return to mixed rotational crop and livestock farming. Instead of artificial fertilizers, livestock and cover crops, like clover and legumes, are used to feed and heal the soil. Mobs of cattle and sheep eat and trample the crop after harvest to a thatch and subdue regrowth, and eat the arable weeds when the field is returned to grass in the rotation. It turns out the old rules of the field mostly still hold true—we just need to minimize or end plowing. Ironically, the best new sustainable "technologies" for making exhausted crop-growing soils healthy and fertile again are cows and sheep. That is why the uplands were historically the livestock nursery for lowland areas; huge numbers of sheep and cattle were needed to make plant-based farming possible, and it made sense to produce many of these on the lowest value marginal land.

~

Back at the house, we are scolded by Helen for getting wet through and coming straight in and making a mess. Tom has handfuls of fir cones and drops them in the doorway as we arrive, which doesn't help matters. Isaac strips to his undies and disappears upstairs to change. "Take your clothes with you to the wash," shouts Helen. I get my socks off and sit in the oldest chair. Our kitchen smells amazing. Helen has made a chicken-and-leek pie and potatoes. I tell her about the fencing and the morning's jobs, and she humors me, pretending to listen, but is preoccupied with her own work: the mail on the worktop, invoices spread out. Her laptop is open. She has spent the morning going over our medicine book because we have an official inspection next week. She does the countless unheralded jobs that keep our home and farm going. And sometimes she looks as if she wants to strangle me or scream at me for leaving her in the house with four kids and a million things to worry about. At other times I see her smiling as our children play in the fields or the beck, or when they are helping us in the sheep pens or the barn, and it seems that she loves this life too. She has fire in her and fights each day for our family. I can only do what I do because Helen devotes her waking hours to supporting the farm and caring for our family. She keeps everything going behind the scenes. She has always been the strong one when I am falling apart.

~

I'm not sure I am much good at being a farmer. It is over-whelming. I can't get everything done, let alone done well. I am always juggling responsibilities to pay the bills. Often, I get things wrong. The farm makes almost no money, and whatever money it does make, it devours. I am terrified of going broke, of letting it all fall apart. I know now why my father looked beaten at times. The list of jobs I haven't got done yet grows each month. I could fill two lifetimes mending our fences and walls, laying our hedges, curing every ailing sheep, and doing all the work needed on our flock to improve it. I could spend vast sums of money we don't have on planting trees and creating more wild areas. I could spend even more trying to breed a great herd of pedigree cattle and peerless flocks of sheep. The truth is, a farm swallows you up, takes everything you have, and then asks for more. It is also an exercise in humility: you can't do it alone.

The biggest lesson I have learned is that the whole idea of the heroic individual "farmer" is a bit of a macho-male myth. It takes a village to make a good farm work. A lot of this work is done by women. My wife, mother, and grand-mother have all supported this farm, built their lives around it, and suffered the men obsessed with it. But it goes far be-yond our family. We rely on so many people who know how to sustain this landscape the traditional way—who can use the old tools to manage hedges, coppice woodland, or grow and plant healthy trees. We turn for advice to good stock-men and -women who know about cattle and sheep—when to move them and how, and how to breed them hardy and for local conditions. We need these skilled and thoughtful farming people now more than ever.

I've come to realize that we also need a small army of naturalists to help us play our part in the restoration of the countryside. There is more to understand about the ecology of a farm than any farmer can reasonably be expected to know. Admitting I couldn't do or know everything was scary to begin with. I worried I might not be as strong or as wise as a farmer is meant to be. But the more I have shared and accepted help, the more our farm has become a community. Several ecologists now have a role in our farming—some from government agencies, some as supportive friends, and some paid for by us to help us understand what we have and why it matters. The pooling of knowledge is transforming our understanding of our land and our valley. This coming together of old and new knowledge on our land makes our farming the most exciting and rewarding it has ever been. Life is richer for all these things we are learning. And the more we share our farm, the more it feels like it matters. I am no longer turning away from the world.

Our fields are busier than they used to be in the 1980s and 1990s, when they were sometimes lonely and quiet places, as layers were stripped away and work and people were lost. This valley is now increasingly a place of regeneration—where work, good skilled work, is being done again. There is always someone else here working, or learning, or helping us. Restoring the countryside isn't about destroying old communities and traditional ways of life, or at least it shouldn't be. It is about building strong new rural communities that respect both the old and the new.

Schools come to our farm and use our fields as classrooms for learning about farming, food, and nature. Seeing children watch a lamb be born, touch a fleece, or follow how food is produced from field to fork is a joy. They gasp at caddis fly larvae in a petri dish, scooped from the becks. They scour for woodlice in rotten tree logs. They rummage through the understory of our wilder places for frogs and toads. They gallop around counting wildflowers in our meadows. And they help us plant hedgerows or stick willows in the riverbanks. During their breaks they revel in the freedom of the fields, running around yelling, filling our land with a wild happiness and freedom.

One day a class came from a distant town, and we were told that one little boy was having a rough time at home. He could barely speak, was as gray as concrete, and flinched each time an adult moved near him. At lunchtime, his teacher and I took him to collect the eggs from the little waist-high mobile hen hut that stands on the grass by the barn. We spoke gently about the hens clucking around the yard, and how they come back to the nest boxes to lay their eggs in the hay. He reached in and lifted out the warm eggs, and his face glowed for a second with the sheer simple joy of it. As the day went on, he seemed to get the color back in his cheeks and find his voice and a little more confidence. He loved seeing the sheepdogs work our sheep. And when he said goodbye, he smiled at us as if he really meant it. His teacher said he had had an amazing day, his best for a long time. When he left, I cried with frustration and sadness.

~

After lunch my daughters come with me because I need them on the lanes to move one of our flocks to the pens. Some of the lambs have dirty backsides and need worming. The flock is up past the village on an "allotment," some rough pasture ground enclosed a long time ago from common land. Molly stays at the lane end and sits on the grassy bank sunning herself while Bea and I go for the sheep up the road. Bea stops the cars as I get the flock out of the field. We chase them down the road, with our five dogs behind them, and as they reach Molly she turns them down the lane and we head for home between the willow trees. The girls swish stripped ash wands from the hedges.

"What kind of tree is that?" I ask.

"A rowan . . . Easy," they reply.

"Who farms this field?"

"Peter Lightfoot."

"What do you think of that Herdwick tup?"

"OK. But it's a bit light-skinned for you, Dad."

They are proud of their farming culture, having shown and sold sheep since they were small.

~

I have tried to teach my children the things I was taught, like how to "know sheep," being able to recognize who they belong to, and to weigh up their quality in a few seconds. They understand the style and character of our flock and the others around us, and they also have sheep of their own. Molly lost her eldest ewe two winters ago. And, because she worked for me at lambing time, I let her choose a young ewe as a replacement when a hundred or more sheep were in the barn. They would have looked identical to most people. A few minutes later she chose the best one and dropped a cool teenage smile at me on the way out of the shed. She knew it was the best, and one I had hoped not to give away, but the loss was lessened because I was proud that she was knowledgeable enough to pick it. Not many shepherds would have chosen so well.

When Tom was born, I purchased a ewe lamb at a sale from our neighbor and rival Jean Wilson for £500 to start his flock. She wrote him a card saying she hoped he would grow up to beat me at the shows with the lambs this ewe would breed. She gave him his "luck money" (£20 back on the purchase price to wish him well). In the years to come he will, no doubt, hold sheep of his own at shows as well as my girls can (they are a match for most of the shepherds).

~

I also want my children to know how fortunate we are to live in this raggedy old farming valley that is so full of nature. To see—as I did not when I was young—the bigger picture beyond our traditional horizons and think of our farm in its wider setting; to appreciate that no farm is an island, but part of a wider ecosystem, a valley, a river catchment, an interconnected world. I point out the chimney sweepers, the little black moths that are being kicked up by the sheep, others with little spots called ringlets, and meadow browns, the larger brown-and-red ones at the verges. I show them the lavender-blue bugle flowers that rise up every few inches from the turf, the simple little yellow flowers of tormentil, and the moss on the damp side of the oak-tree trunks. As we pass through our open gate and across the pastures, I lift some brittle papier-mâché cow muck and my girls look at me like I am weird. It crumbles in my hands but is riddled with life—fat gray grubs, little black dung beetles, tiny turquoise beetles, and insect shells that sparkle in the sunshine.

~

The ewes flick their ears and lower their heads to escape the flies that are gathering above them. The oldest ewes are leading the flock, their lambs trotting behind them. The dogs weave from side to side, pushing them along, tongues flopping out of their mouths. I try to explain what I am doing, why I care about muck, and beetles, and peat bogs, and fungi, and worms. I try to explain that the future of the farm lies in replicating as best we can the habitats (and natural processes) of this once-wild valley, thousands of years ago.

"But what was it like?" Molly asks.

I have to admit that for most of my life I didn't really know. My grandfather acted as if our valley was what it had always been and always would be. But now ecologists tell us that it was once much more wooded and had species in it that I have never seen—creatures such as corncrakes, polecats, pine martens, lynx, wild boar, bison, wild cattle, beavers, bears, and wolves—and species that no longer exist here, like woolly mammoths and rhinos, bog elk and cave lions. Ecologists aren't completely certain how it all worked, because there is no description, no photograph, no film, to help us imagine that past. Some believe this land was a closed canopy of mature trees, the deep dark wood of European fairy tales like "Little Red Riding Hood," but growing evidence suggests that that is mistaken, or at least overly simplistic. Humans have been part of it, shaping it, for tens, perhaps hundreds of thousands of years. We came to places like this as hunter-gatherers before the last ice age, and before the woodland reclaimed it, following herds of reindeer and other herbivores through such landscapes, as people still do in the far north.

The idea of it all being forested is more likely to be a projection based on what happens now when we abandon woodland to do its own thing, rather than what probably happened in the prehistoric wilderness when large herds of herbivores were moved around by large carnivores. Wild systems are full of what ecologists call "dynamism" and "disturbance." Large and small herbivores graze trees and scrub and smash it down, creating clearings and grasslands. We tend to think of habitats as being static and fixed, but in nature all habitats are forced to change constantly by grazing, storms, disease, smashing, trampling, rotting, death, and decay. And all of these processes are necessary because they create unique niches. A healthy ecosystem is in a state of perpetual motion. Real wilderness never looked like a genteel English wood. It was messy, and the three main habitat types—woodland, scrub, and meadow clearings— were moving around in an ever-shifting, swirling dance of birth, life, and death.

Molly says it sounds a bit like a documentary we watched recently about the Serengeti. She's right, I think. Our traditional farming systems can evolve to provide a similarly shifting patchwork of habitats (using the native breeds and the craft skills of the people of the countryside): meadows and pastures that resemble the woodland clearings of the past, with different intensities and types of grazing, and thick hedges like those we are passing now with the flock, full of birds that think they are in a thorny wild woodland edge. As we add more of those vital habitats and processes, like wilder rivers and ponds, willowy and thorny scrub, the species we once lost will begin to return. Ironically, the best farming in Europe for wildlife is in the least developed, or "backward," places like Romania and Hungary.

~

The sheep in front of us have reached a beck. They bunch up, hesitating, until one jumps at the narrowest point and then the rest follow. Soon we turn into our gateway and down to the pens where the sheep are gathered in. The girls grab the dirty ones, and we hold them firm as I clean the filth from their back ends with a pair of hand shears. One lamb has some maggots on a damp patch of its wool; we flick them off and apply Battles Maggot Oil. The lamb shivers and looks relieved. Half an hour later we are done. We chase the flock through the pasture above the pens and into a fresh field. A redstart bobs along in front of the flock. Then another, and another, until five little red tails are flitting along by the sheep from one thorn tree to the next. Their tails flash with each wingbeat, little triangular wedges of freshly cut mahogany. The air is heavy with mint, trampled by sheep feet. And then they are in the new field, heads down, grazing, or up, calling for their lambs. I tell my daughters to set off for home on their own and they race each other down the steep bank covered in white, pink, and yellow flowers. I shout "thank you" after them, but they can't hear me. I must go back to the cow and calf to see that all is well.

~

The meadow looks entirely different now from the way it did at dawn. These fields never look the same twice. I am beginning to learn how to read our meadows properly, and I am proud of them. It is a new kind of pride, to add to my knowledge of my sheep and cattle. I have always seen their beauty, their swaying waves of color, but I didn't see, or understand, their wild biodiversity. My grandfather taught me about the grasses, but he never mentioned the wildflowers. My grandmother and mother would talk about the prettiest flowers in our meadows but were often unsure what their names were and would promise to check *The Observer's Book of British Wild Flowers* that lived in the telephone drawer.

As the months passed after my father's death, I became more aware of how little I really knew about the nature of our valley, so I paid a botanist to do a plant survey. Within seconds of him walking into our best hay meadow, I could see that he knew how to read it in ways that I could not. He splayed a collection of flowers in his hand and talked me through them, pleased that I was interested. After an hour or two it became clear that our meadows were much better, more biodiverse, than we'd imagined. Our fields weren't ruined—and it wasn't too late. He was finding obscure plants that he had to look up in his books—plants he'd rarely seen before or had not expected to find here. By the end of the first day, he was sunburnt and glowing with enthusiasm. He had recorded over ninety species in the first meadow. Some modern intensively farmed fields of grass now have just four or five species in them, sometimes only one species. He pointed out plants that I had never noticed before in my fields and told me their stories: arctic eyebright, confused eyebright, marsh pennywort. And he gave names to plants I had known my whole life but still had remained ignorant about, such as wood cranesbill, marsh cinquefoil, melancholy thistle, ragged robin, harebell, flea sedge, marsh ragwort. He found close to two hundred species of plants and grasses on the farm, many of them on the endangered "red list." He also realized that half a dozen important species weren't in our meadows, so we set to work reintroducing them with 6,500 tiny plug plants, each one dug in by hand.

The botanist taught me something else important, that "nature" isn't something that just lives around my fields and on the scruffy edges—it lives *within* the fields as well, in the soil and in the sward. Rare plants are wonders, but we also need lots of common plants too. A field with twenty species of flowers and grasses in it may not be as ecologically pristine as a wild beaver-made meadow, or a bison-grazed woodland clearing, but that doesn't render it worthless. It is much richer and better for nature than a flowerless silage field. Some fairly common plants like red or white clover, or daisies, or buttercups, provide a lot of food for insects like bees.

The idea that land must be either perfectly wild or perfectly efficient and sterile is unwise and blinding; it is a false and unsustainable simplification. When we despair and reduce our worldview to black and white—"farming" is bad; "nature" is good—we lose sight of vital distinctions and nuances. We make every farmer who isn't a saint a villain. We miss the actual complexities of farming, the vast spectrum between those two extremes and the massive scope for nature-friendly farming that exists there. Some quite modest farming changes at scale could be revolutionary. Whatever the farming system, there are ways to nudge things to get lots of nature back into it. Just as marginal farming gains over the past fifty years incrementally squeezed nature out of our fields, the process can be reversed.

~

We won't win any prizes for our farming or make a lot of money from it. And it won't "feed the world" (the proud boast of the least sustainable farming systems), though we can contribute to that. To some farmers at the more intensive and industrial end of the spectrum, our ways will seem like a nostalgic daydream. They will make different choices: they will feed more people more cheaply than we will. Every farm cannot and shouldn't be exactly like ours—that's OK. To have a robust and resilient food system we need many different kinds of farmers. Diversity is a strength in farming, as in so much else. But whatever the type, farms can be vastly better for nature with some tweaks, and by farmers being supported to shift away from damaging practices.

~

It is late afternoon now, and the shadows on the southern face of the fell are lengthening. I have seen that the calf is fine and full of milk, its coat licked into curls by its fussing mother. I walk across our land and feel a sense of joy at the way it is changing. My simple rule is that wherever you stand on our farm you should never be more than three hundred yards from another different kind of valuable habitat. I want a farm full of birdsong, insects, animals, and beautiful plants and trees. It should run overwhelmingly on sunlight—not on fossil fuels. We are shifting to using fewer drugs and chemicals, and less bought-in feed. We use almost no pesticides, and I hope it can soon be none. Already we are off-grid for electricity, with solar panels instead of utility lines, and the next stage is a mini hydro or wind turbine. Recently we were carbon-audited and were found to be trapping much more carbon than we were using, or releasing, and I think we can trap much more.

None of this is at odds with breeding a great flock of sheep or a herd of cattle. The fertilizers, medicines, pesticides, fuels, feeds, tractors, and machinery that we once bought that made our farm lose money have turned out to be the very things that did all the damage. We are moving toward a farming system that produces food with the minimum of artificial inputs. Sadly it means earning money away from the farm when we have to. This new way of thinking about our farm doesn't replace my original dreams or sense of identity; it just adds layers of interest to it.

In truth, we are in part returning to an older type of

farming life—one of sweat, blood, and hard work. The work becomes seasonal and physical once again and revolves around being there and seeing things and getting our hands dirty. It isn't a recipe for an easy life. Farmers in the past worked extremely hard and had to be eternally vigilant. It was a tough old game and doesn't fit with any economic principle of minimizing work and maximizing productivity. A farmer's focus shouldn't be about spending the minimum amount of time on the land. We should be present on the land constantly, understanding and utilizing its rhythms and processes better, and caring for it in a more hands-on way.

Our new farming life means we have less power and control. We are engaged once more in the old tussle with nature and we often lose. Last winter I tried to minimize the medicines and antibiotics I used on the animals, and when a severe blizzard came late in the spring my sheep suddenly became ill and coped very badly with the howling wind and biting cold. They had, unbeknownst to me, been carrying flukes in their livers—a parasite that uses a tiny freshwater snail as its host for part of its life cycle and then is eaten by sheep on the grass and damages their intestines and livers. This whole process is invisible, but when the long winter put my sheep under severe pressure—when they were lean and tired—they began to struggle, pneumonia set in, and several became ill and died. I knew that if I had treated them with one of the powerful drugs for killing flukes we might have avoided this. Sometimes stepping away from these things feels like naivety and takes a heavy toll. If the past had been easy we would not have abandoned it quite so readily. When your children are ill, you want the best medicines, and my attitude is the same for my animals. But we will try to keep only native breed animals that are adapted to our land, and select them to manage with the minimum use of chemicals and drugs.

~

My father drilled into me, in the months before he died, that I didn't have to fight the whole world. In his last fifteen years he attained a kind of wisdom that perhaps he hadn't always had (or that I hadn't always valued). He'd learned that it was all right to step back sometimes and take a breath, and do something different if it paid better, or made more sense. It is OK to admit that you don't know, or that you might have been wrong previously. That pragmatism is perhaps my father's most useful legacy. He and my grandfather didn't do all the things I am doing, because in their eras it wasn't expected of them. I can respect their good sense, and their work, without copying their every move religiously. They responded to the ages they lived and worked in, and so must I.

~

It took me a long while to accept the full extent of Lucy's plans for river conservation. Her big idea was to shift the stream that ran through our best hay meadow to a more natural channel a few hundred yards away, in order to create new ponds and wetland areas. It seemed to me that it would take too much land and spoil our meadow, and undo too much of the work of the past. So initially I said no, and we settled on some less ambitious changes. But over the years I have learned more about wilder rivers and accepted that we can work around them—and I changed my mind. I asked Helen, and my children, and they said we should do it because it was right.

This summer the diggers came: three men from a local construction company with a digger, a dumper, and various other machines. They are a bit skeptical about using their machines to make a field "worse," and think this plan is the kind of lunacy only college-educated people could come up with. But they have got into the spirit of it and have dug scrapes and ponds, and a twisting and turning channel for the river (the route identified by satellite-modeling of the flooding). Then they filled in the old straightened channel. Now that's been done, they have begun proudly to create the ponds and scrapes that are even more ambitious than the plans we had. Finally, it will all be fenced off for occasional cattle grazing so that the remaining meadow can work as a lambing pasture, and newborn lambs won't be able to get into the becks and drown.

We are making plans to plant willow strips and alder and hazel on the riverbanks. Helen says I like "planting beaver food" and that before we are old these saplings may be chewed off by those returning furry river engineers. Time will tell. I once thought such reintroductions were madness, but my views have softened. The more I learn, the more it makes sense for beavers to manage the wettest bits of our valley someday instead of me with an ax and a spade. My cattle and sheep can graze around a wilder river.

I am taking my father's advice to adapt to change whenever I can and become open to new possibilities. It is not my job to fight every little change in the world.

~

When we come out of the house after supper, the ewes and lambs are calling to one another across the valley. I take Tom and Isaac to train our youngest sheepdogs, Bess and Nell. At the barn, they jump excitedly and nuzzle us, looking for praise, as soon as I open the stable door. Isaac has to holler at them and whack them away with his hand until they respect him. Tom clings to the fuel tank of the quad bike between his knees, bent forward as they jump onto the bike behind him. Then, as we head down the lane, they jump off, race ahead, and scrap with each other, tumbling down the bank between the thistles.

Isaac is the kindest, brightest, and most loyal boy any father could hope for. He is a bookworm, and as we go he tells me tales from the Norse myths he has been reading. His farming dream fills me with pride, hope, and fear. Pride and hope because I would, of course, love one or more of my children to be farming here one day in this place that has become my life. I love the idea of this old farm continuing on, and that one or more of them might feel something like the love I feel for this valley, and have the same sense of purpose it has given me. But I don't want them to feel trapped in their father's dreams. I fear for them too, because it is at times a crushingly hard way of life, and the economics of what we do are terrible.

We train the young dogs on some old ewes in the bottom meadow, sending them out around the field's edge to bring back the sheep to a pen made of hurdles in the middle of the field. Bess is strong and keen. She loves work. The ewes have learned that they cannot escape her and put themselves in the pen without much resistance. Bess looks at me, frustrated that such feeble sheep have been her lot. Her sister Nell is timid, and it is all I can do to keep her reassured. But after a little while she is turning them and has a certain silky style. I regret keeping them both, because Bess is a bit of a thug and seems to have affected the confidence of her sister. I will have to give Nell more praise and time away from the rest as I try to make a sheepdog of her. Instead of rushing around, chasing our tails, I'd like to get our farm on an even keel, so we can breathe, slow down, and devote more time to jobs like training the sheepdogs and enjoying this amazing place we call home.

~

We go to see my tup hoggs, the year-old males we keep for selling as tups (rams). They spend their summer up on a plain little field belonging to a man called Barrie who, Isaac insists, looks like Father Christmas. Something about "Barrie's Field," perhaps its peaty acidic soil, helps darken our Herdwick fleeces to a deep slate-blue, the ideal color for selling them in the autumn. Their white heads and legs have "cleaned out" and brightened in recent weeks, losing their black lamb color, and now their head and legs fill my keen eyes with a whiteness like snow. Their noses are wrinkling and bulking out, with testosterone kicking in as they reach sexual maturity. Their legs are becoming sturdy, their bones thickening, and behind their heads their woolen necks are silvering like manes. Their pale horns have curled around in recent weeks, twisting about their ears, and often close to their eyes. There is a pecking order, and the strongest among them strut about like kings.

I love my flock of sheep; they remain my pride and joy and, I suspect, they always will be. I work with Isaac on his sheep-judging skills, helping him to see the little things that set a great sheep apart from a field full of good ones. He is just learning the basics, but I think he will be OK. Last autumn I was unable to go to our shepherd's meet to show our sheep; my other work had swallowed me up. I didn't have time to make the sheep ready, so instead Isaac went with my mother, and I stayed away, frustrated not to take part in a day I love so much. That night Isaac returned proudly wielding a silver plate he had won in the "young handlers" competition—an award for the young shepherd who had displayed the best ability to present a sheep to the judges. "But you didn't take a sheep," I blurted out. "Oh, that was OK," he said. "Jean Wilson lent me her tup lamb, and it stood itself up real proud and proper. I thought I'd win." Jean loved that—sending a message home to me that if you held one of *her* sheep, you win the silverware.

~

I was brought up to think in a certain way about what made a "good farmer." The best farmers had the best cattle, sheep, or pigs—they were great stockmen. They had fields brimming with amazing crops, and they worked hard. To me, this worldview felt like it had never changed and never would. But I can see clearly now, as I look back at the farming lives of both my father and my grandfather, that what it means to be a good farmer evolves from generation to generation.

I grew up understanding that a farm was a piece of property, a private thing owned by someone, a family's entire wealth, or else a tangled legacy of debt and obligations. It was above all a place of work—work that defined people and gave them purpose. It was also a business, a commercial enterprise, producing food to pay the bills and feed other people. Feeding people is a noble endeavor, but is now often taken for granted despite its importance. Human societies are laden with risk—get farming even slightly wrong and people begin to go hungry, the poor first; get it badly wrong and millions starve.

A farm was also a home, a taproot for a family, deep in the ground. It was still a home even if it didn't make any money, which explained why farmers would hold on long after they should have quit, a place layered with histories, stories, and memories, like sacred ground. At the same time a farm was part of a wider cultural, social, and economic system, a network of families and communities doing the same work, devoted to the same breeds, plants, or practices.

An old farmer once told me that it felt as if he had ceased to exist for eighteen months after he lost his cattle and sheep to the foot-and-mouth disease epidemic in 2001. It was not just that his sense of identity was bound up in his work with his animals—and the sales, shows, and other gatherings around them—it was that all of his friendships and relationships were tied to it as well, so that his links to his community disappeared when his animals went.

A farm always represented a rural dream of being independent and free, making a mark in a place, rather than being swept up in a vast exodus of strangers to a city. All this is still true, and yet I now understand that a farm is a once-wild place that was tamed for our purposes, part of an ecosystem that has often become broken or impoverished. Our fields are the coalface where we as a species meet the natural world, where our politics, our diet, and our shopping choices shape the land, the wilder world around it, and even the climate. And we have often done great damage.

I have come to understand that farming, even in the traditional ways, always has costs for the natural world—it is usually a downgrade on what might be there if humans weren't. But once we accept that, we can also see that good farmers do more than produce commodities: through benign inefficiency or good stewardship, their farms can allow a great many wild things to live in and around them; holding water that would otherwise flow off the land and flood villages, towns, and cities; and storing carbon that would otherwise alter the global climate. A good farm has a public value that transcends the pitiful price the farmer gets paid for his products. One of my shepherd friends has over a million people a year walk through his farmyard on their way to and from the mountain he grazes. They come from all around the world to admire a beautiful, rugged, half-wild landscape of fellsides and becks and crafted drystone walls and old stone houses and barns.

I have come to understand that even good farmers cannot single-handedly determine the fate of their farms. They have to rely on the shopping and voting choices of the rest of us to support and protect nature-friendly, sustainable agriculture. They need government spending and trade policies to recognize that sound farming is a "public good," a thing that needs encouraging and protecting.

The marginalization of farming in our national politics and culture is a tragedy because this is all about the kind of country we want to live in: a second-rate version of the broken American Midwest, or a land that reflects our own values, history, aspirations, and nature. We won't become a country of good farming by simply drifting along as we have been, letting big business become ever more powerful and insisting on ever cheaper food in the supermarkets and shops.

We have to take food much more seriously, viewing it not as a technical problem to be solved, but instead as something important in its own right, that enriches life. We need to think about how food is produced, and how our choices play out in fields somewhere. We are all responsible for the new industrial-style farming. We let it happen because we thought we wanted the sort of future it promised us. Now, if we want a different kind of future, we need to make some difficult decisions to make that happen.

We have spent too long listening to economists. They said we shouldn't worry about local food because we had secure global supply chains. But even if that dubious claim were true (the world is much more volatile and vulnerable to human and natural crises than they admit), that isn't why local food matters. We need local farming so that we can understand it and engage with it, and shape it to our values. That means a significant share of our nutrition should be produced locally so we can see it, participate in it, and question and challenge it when we need to. Food production is too important to be pushed out of sight and out of mind. Foodstuffs from anonymous distant global sources are rarely subject to our rules and regulations on welfare, environment, or hygiene, or produced in line with our values. We are so used to being disconnected from the fields and the people that feed us that we have forgotten how totally weird this is from a historic perspective. We should not be strangers to the fields that feed us. It is bad for us to get too far from the soil and the elements, and the tough realities that sustain us. There is now a mountain of evidence that people are healthier in body and mind when they do physical work and spend time outdoors and in contact with the natural world. The modern aspiration of lifting us up and away from elemental concerns and work on the land has turned out to be exactly the opposite of what we really need.

We need to keep unsustainably produced food out of our shops and markets; it cannot be allowed to undercut nature-friendly, high-welfare farming. We must continue to produce lots of our food on British soil in order to avoid importing more from sterile, ruined landscapes like those of the American Midwest, or from land being cleared of pristine ecosystems in places like Indonesia and the Amazon. In such areas farming has less regulation, less control, and less oversight by us.

Above all, we need farmed and wild landscapes that balance our complex needs better. We should gradually limit use of some of the technological tools that have been changing farming over the past half-century, so that methods based on mixed farming and rotation can be reestablished.

Take fossil fuel–based chemicals away (over time) and farmers won't have to be told to be mixed or rotational, they will return to it themselves. If we can encourage more diverse farm habitats, rotational grazing, and other practices that mimic natural processes, we will transform rural Britain. We don't need governments to micromanage farms (there is a long history of failure in that, for example in the old Soviet Union), we simply need to create the right systems for farmers to do these things viably.

The old social contract between farmers and society is now stretched to the breaking point. We need a new deal, a new understanding, a new system, that brings farming and ecology together. And that requires dialogue, realism, trust, and changing our behavior as both farmers and consumers, and a willingness to pay the real price (in the shops or through our taxes) of food and good farming to make things as good as they should be. Some of the solutions are small and individual, but others require big political and structural changes. We have to flex our political muscles in our millions to create a politics that sees the land and what happens on it as being at the heart of building a more just and decent country.

~

Tom is fast asleep on my knee by the time we get home, heavily slumped in front of me on the quad bike, with my arm around him. Helen comes out of the house and rolls her eyes. She helps me lift him off and carries him in. There are a great many things that are unfair in our rather old-fashioned household, as I am rarely indoors to lend a hand or do my share, but every evening I put Tom to bed. He is a little bull of a boy, all will and determination and untamed energy. Going to sleep isn't in his plan; he usually twists and turns and fights it. He loves to be read Jane Pilgrim's stories of Blackberry Farm that were once mine and my father's before me. But tonight he is out like a light. I stroke his hair and wonder what he will make of us years from now. I hope I will have done enough.

~

What will our descendants say of us, years from now? How will we be judged? Will they stand in the dust of a scorched and hostile world, surrounded by the ruins of all that exists today, and think that we, who could have saved the earth, were thoughtless vandals, too selfish or too stupid to turn back? Will the future know us as the generation that pushed everything too far, on whose watch the world began to fall apart, who had so little courage and wisdom that we turned away from our responsibilities? Or will our descendants lie in the cool green light of the oak trees we planted and be proud of us, the generation that pulled things back from the abyss, the generation that was brave enough to face up to our own flaws, big enough to overlook our differences and work together, and wise enough to see that life was about more than shop-bought things, a generation that rose above itself to build a better and more just world?

This is our choice.

We are at a fork in the road.

A prudent gambler would not bet his house on our virtue, because the odds say we will fail. There are a million reasons to believe that we are not big enough, brave enough, or wise enough to do anything so grand and idealistic as stop the damage we are doing. We are choking to death on our own freedoms. The merest mention that we might buy less, or give anything up, and we squeal like pigs pushed away from the trough. The world of human beings is often ugly, selfish, and mean, and we are easily misled and divided. And yet, despite everything, I believe we, you and I, each in our own ways, can do the things that are necessary.

~

A few weeks ago, we planted the twelve thousandth sapling on our land. We have probably tripled the number of trees on our farm—and we hope to plant many more in the years to come, to run more strips of scrub and woodland through our land. I have challenged myself to plant, at least, a tree for every day until I die. That won't create a forest, but it will be something. I want to create a farm full of shelter, and patchiness, and shade, with leaf matter everywhere returned to the soil. I want to make our farm even better for wintering birds to come to for berries and fruit. Our farm solves nothing on its own; it is just the little place where we make our start. But all of us together can transform our landscapes one little act at a time.

There is something about planting trees that feels good. If you have done it well, it will outlast you and leave the world a little richer and more beautiful because of your efforts. Planting a tree means you believe in, and care about, a world that will be there after you are gone. It means you have thought about more than yourself, and that you can imagine a future beyond your own life span, and you care about that future. My father mended the broken gates and walls before he died because he believed that this farm going on mattered, and through it so did his modest life. I believe that too, and plant trees with the same sense of purpose.

They say that old wheelwrights planted, felled, and stored apple trees in a three-generation cycle, so that their grandsons would have sufficient matured trees, and dried wood of the right kind, from which to make the hard wheel hubs they needed. We need to live like that again, thinking longer term and with more humility. Maybe my descendants will make something from these trees we are planting now. But, less grandly than that, I hope my children and grandchildren will climb trees and build dens, catch fish and roam free the way we once did. I hope they will love our flocks and our culture and cherish the wild things. I am tired of absolutes and extremes and the angriness of this age. We need more kindness, compromise, and balance.

Everything good we have done on our farm has come from people finding ways to bridge the historic animosity between farmers and ecologists. The old freedom to do what we like as if each farm were an island, just a business, separate from all others, is now problematic. One appropriately chosen field of maize in a landscape may not be a major problem for soil erosion; all of the fields being planted with maize might be an ecological disaster. Ecology is bigger than one field and one farm. We need to work across many farms, and many valleys. If we know anything about ecosystems it is that they span whole landscapes, from sea to mountaintop, from north to south and east to west and back, and encompass farms, valleys, regions, and countries, even continents.

The cuckoos that call from our oak trees for two months each spring come here from sub-Saharan Africa. They need many safe places to live and feed, and to be able to pass safely over as they migrate. The same applies to the swallows, swifts, stonechats, spotted flycatchers, redstarts, and many others. My farm is intricately linked to faraway places that I'll never visit, and over which I have no influence. The fieldfares and redwings I love in my hedgerows all winter disappear one day each spring and fly north to the Arctic tundra and forests of Scandinavia and northern Russia. They say the ravens that pass over our farm—and which strip the flesh from the dead sheep on our fell—sometimes fly northwest to gather and find mates, and to escape the cold spells on the coasts of other countries, like Denmark or

Norway. But even on a more local scale, wild things move around between valleys and areas constantly. The curlews, oystercatchers, and lapwings that grace our valley move between our fields and the mudflats of tidal estuaries in places like the Solway Firth and Morecambe Bay. The otters that leave their spraint (dung) marks on the rocks below my farmhouse have much larger territories and move up- and downstream of our farm, covering many miles in a night. And the salmon and sea trout that flash silver in the pools of our becks spend much of their lives far out at sea dodging trawler nets and the mouths of killer whales. Our little farm is part of a very big world.

~

We cut across the fields on the quad bike, through the gateways, with Bea jumping off to open each gate. We are almost on the floor of the valley now and fingers of sunlight stretch across the fells to the northwest. Shadows lengthen. I have not seen our stock tups yet today, the most valuable sheep on the farm. They are the source of new blood for the flock and I must see they are all right before nightfall. Isaac has stayed behind to watch TV until bedtime, but my Bea was keen to come.

We are heading for the marshy land of the floodplain. Carved apart with countless little becks and old drainage channels, it is rough grazing land and later in the summer will be waist-deep in creamy meadowsweet flowers. It is here that the tups are meant to be. At dawn and dusk the valley bottom feels a little primeval, with the cattle and roe deer often grazing in a sea of mist, while herons pass to and fro from the places where they fish and catch frogs and roost. For centuries it was drained heavily to turn it from a bog into more usable fields, with giant channels as deep as eight or ten feet. It took Herculean efforts and a lot of men to keep it drained. The final act of "improvement" took place in the 1980s with powerful machines when the "Water Board" straightened the beck across our land. They lined the river on either side with posts and wooden planks until it looked like a tiny canal. That a utility company had the money or inclination to do this in a remote Lake District valley out of little more than a sense of engineering zeal and a theology about straight rivers already seems surreal. But scarcely a field in England hasn't been "improved" in one way or another in the past century.

I have neither the staff nor money, nor the will, to drain this place. Hairy old cows and Herdwick sheep don't need ramrod-straight rivers or perfect green sward. So, without any great master plan or fanfare, the valley bottom has, year on year, been reverting to a wilder place. The straightened river has eventually eroded the banks put in place to hold it, and now threatens to break loose. The canal planks are rotting. The gutters have silted up and are slowly filling in, with sludge and pondweed, and otters and herons hunt frogs in them.

As we travel into the valley bottom, I see around me on

all sides an ancient working landscape that still lives and breathes, but also with twenty years of changes written across its surface.

I see ancient oak woodland above us trying to regenerate. Little mountain ash trees are sprouting up all over the wilding fell, trying to beat the deer. The vegetation is growing denser and deeper, with alder and thorny scrub creeping up the gills. The floodplain is half-abandoned and half-wild. The valley has become much shaggier and wilder than it ever was in my childhood, with far fewer sheep dotted around. Some of my neighbors are confused or angry about that, while others are adapting, keeping more cattle or finding other ways to earn a living from their land.

I see farmers starting to work together to make this place even better, finding ways to farm around wilder rivers. Miles of hedges are being laid once more, drystone walls rebuilt, and old stone barns and field houses restored. I see river corridors fenced off and ponds dug; the blanket peat bog on our common land has been restored. Wildflower meadows liberated from artificial fertilizers and pesticides are now shimmering with clouds of insects, butterflies, moths, and birds.

And I see other people in our community who aren't farmers also planting trees and hedges, or creating wetlands, or helping to coordinate our efforts. These things bring separate worlds together, and the old "us" and "them" divide is fading. There is a love of this place that unites us all.

~

Sometimes I do not know what to make of all the changes, so I simply watch and learn from what emerges. I am not arrogant enough to think I have all the answers. These landscapes are being shaped by many people and many ideas, as they always have been.

Still, through all of this runs a thread of continuity with everything that has come before, as most of the hefted flocks of sheep still follow the same movements between fell and pasture they have always done. We still work together on our commons with sheepdogs and gather and show our animals at shepherds' meets. These valleys are full of keen and smart young men and women who love its traditions and the work of the old fell farms. They are desperate to play a role in this way of life, and build their future in it, just as I was.

This valley may remain unloved by both die-hard production-focused farmers ("It is cute, but just a lifestyle choice") and extreme wilderness-loving ecologists ("Please could you just disappear because we'd rather the uplands were forested"), but to me it represents a beautiful compromise, and it is improving all the time, as we learn new things and find fresh solutions to its challenges. I am proud of my community both for keeping the old ways going and trying to find new ways to address the desperate problems of our age.

I believe in this landscape and its people.

~

I drive the quad bike quietly along the grass lane by the small beck, through the marsh thistles. The tups are standing on the riverbank. They are all head, shoulders, and horns. I count them and see that each is in good health. My eye always fixes on the largest and proudest of them, called "The Beast." We bought him three years ago, and he has bred many fine sons and daughters. By his side is "The Jedi," who we bought for a record price two years ago. These two woolly gentlemen are shaping our flock, hopefully improving it. They turn and gallop down the field.

As we turn, something ghostly white flashes beyond the dike. I stop the engine.

Time slows. Water trickles over the pebbles between the darker pools, glistening in the last sunlight of the day. In the biggest, darkest pool, clouds of minnows, little trout, swirl around erratically, their bodies scratching scribbles on the skin of the water. The air is alive with the gentle hum of flies.

Bea sits still in front of me. She is wearing a pink T-shirt and shorts, her stubby bare legs on either side of the fuel tank. We wait. I have my arm around her middle and squeeze her tight to let her know I love her and am grateful for her help. She is a fiercely proud and independent girl. She has a broad, open face, a band of freckles across her cheeks, and her hair in a ponytail. She is kind and fun: children flock to her when she enters a room. But with adults she is a rebel, cheeky. Ever since she was small, she has looked to her older sister, Molly, for reassurance and authority, not to her parents. They have always been their own little tribe, with their own allegiances. When I told her as a toddler to do things, she would look to her sister to see whether I should be obeyed. She will defy me, or her mother, if she thinks she is right. Like many of the women in my family, she has spirit, guts, and a keen sense of right and wrong. These moments when she helps me on the farm are the nearest she lets me get to showing her affection. I know that secretly she wants me to be proud of her. And I am, of course, but I'm not sure she knows that yet.

The setting sun casts long shadows from the oak trees across the meadows. The day is nearly done. And then we see it, the ghost-like bird in the field alongside us, fifty feet away. A barn owl.

It seems oblivious to us. There is a tingle of electricity through my daughter's body. We sit in silence and watch it flit backward and forward, like a giant white moth, wings silently caressing the thin twilight air. The owl swings left and right, like a ball rolling from one side of a glass jar to the other, gravity pulling it back down at the end of each arc. Every time it rises, it comes back down the other way. It is so delicate, so fragile, that when it turns away at the end of each hunting arc, it almost disappears. Yet in movement and spirit it is somehow bigger. It fills the field with its presence.

A pitch-black carrion crow, or "dope," as we call them, tries halfheartedly to mob it, but after a quick tumble both birds continue on their ways. The valley is silent, except for the bark of a roe deer. I squeeze my daughter between my knees. She says nothing. She is spellbound.

Such moments are the reward we get for trying to do the right thing on our land. Beauty doesn't pay the bills, of course. It isn't enough on its own, but it makes life better. Country people always knew that before they were made slaves to the gospels of industrial efficiency and consumerism. I am grateful that my grandfather and father taught me that a good life has little to do with money or shop-bought things. Their disdain for the obsessive financial values of the modern world is something I respect more than ever. None of us can escape commercial realities, but we can try to reshape our society to make it fairer, more decent, and kinder. I am sick of 1980s economics bullshit.

Away across the silver-shadowed field, the barn owl tacks from side to side, backward and forward. Then, eye locked on its prey, it folds its wings back and falls like an arrow into the grass. We hold our breath for a few silent seconds that seem to last forever. Then the owl lifts up from the grass, and labors slightly back to a gate stoop, carrying a small, shuddering brown corpse, and we breathe.

There is nothing beyond this. Nothing higher. Nothing more profound than these simple things; nothing that matters more than trying to live our little life on this piece of land.

I hope Bea lives for a hundred more years. I hope she lives a healthy life full of kindness and joy. And maybe when she is an old woman, wherever in the world that might be, she will remember this spot in time when she sat with her father and watched a white owl hunt. A tiny moment of beauty and magic shared. Or, maybe, she will stand in this same place as a farmer, long after I am gone, and remember that I tried my best to look after this land.

This is my inheritance to my children.

This is my love.

Tell them what is happening on the land. Someone has to tell them . . .

When I was young there was cowslips and ragged robin everywhere, and butterflies on the thyme in the rocky crags on the fell. The becks were full of minnows, the pools alive with them, and water boatmen skating on the top . . .

I'm maybe old and stupid, but I like to see them things. But you don't see them anymore. And greed is to blame. Greed. And it will get worse if they don't change things.

Tell them.

Mayson Weir, Dowthwaite Head Farm

Acknowledgments

This book only exists because I have a whole gang of amazing people who make my writing life possible, and to whom I owe a heartfelt thanks:

Thank you to Jim Gill and the rest of my team at United Agents.

Thank you to Helen Conford for commissioning this book and being my first editor, helping to shape it in the early months of its development. Thank you to Stefan McGrath, Ingrid Matts, Penelope Vogler, Jane Robertson, Helen Evans, and all the other Penguins for going above and beyond the normal efforts.

I will always be grateful to my editor Chloe Currens for helping me to make this book what it has become. I feel blessed to have such a brilliant editor who I know will always fight for me and push me to be better. I love being part of the Penguin family—it is an impossible dream-come-true for the teenager who used to marvel at the Penguin Classics on my mum's bookshelves.

Thanks to all those who have helped me write this in countless little ways through our conversations on social media—@herdyshepherd1.

Thank you to all the booksellers, journalists, and festival folk who have supported my writing and hand-sold my books.

Thanks to hundreds of readers and writers for being kind and supportive to me, both face-to-face and by writing. I haven't had time to reply to all letters, but your thoughtful messages have meant a great deal to me.

~

A farming life is full of people who teach you things and help, and I am grateful to all those folk who have done that for me. Thank you to my friends for their support, for keeping it real, despite the hype, and for giving uninvited visitors the wrong directions when they are trying to find my house.

Thanks to Alan Bennet, who is a good farming neighbor and whom I enjoy discussing these things with on the roadside by our fields. Thank you to Peter Lightfoot for keeping me right. Thank you to David Cannon for being a gentleman and the friend he said he would be. Thanks to Joe Weir for being my partner-in-crime with Herdwick tups. Thanks to Richard Woof for all the help and the chainsaw work on those bloody hedges. Thanks to Chris Davidson, Derek Wilson, Scott Wilson, Tom Blease, and Hannah Jackson for covering for me when I am away.

Thank you to Ken Smith for driving me down long, straight Midwestern roads in the United States. And thank you to Renae and Kevin Dietzel for giving me an insight into farming life in Iowa. Although I have written about the bad things in that farming system, it would be remiss not to point out that I have had the good fortune to meet many fine and progressive American farmers and to admire their fighting back—much of the best "regenerative agriculture" thinking is coming out of America. I owe a debt to Greg Judy of Green Pasture Farms for teaching me a great deal about soil and grazing via his YouTube channel.

Our farm has transformed in the past ten years because a lot of people with environmental knowledge spent time with us and helped us to change how we think and how we manage our land. Lucy Butler and Will Cleasby came many years ago from Eden Rivers Trust and began our conversion. Thank you to the current staff of Eden Rivers Trust for carrying on that good work, including Elizabeth, Lev, Tania, and Jenny.

Rob Dixon has helped us understand our land, particularly our wild plants, and has got his hands very dirty laying dikes and planting the missing plug plants that will begin to flower in our meadows in the next two or three years.

Caroline Grindrod has perhaps had as much influence on our farm as anyone in the past few years, through teaching us about soil and grassland management, and she deserves sincere thanks.

Lee Schofield and other colleagues at the RSPB have been an interesting sounding board to exchange views with and have also shaped our thinking. Charlie Burrell and Isabella Tree were kind hosts and guides to what they have done at Knepp Castle—which has also influenced the way I think about our farm. Cain Scrimgeour and Heather-Louise Devey helped us to better understand our moths and bats and made our wildlife fun with their fresh perspective. Thank you to Becky Wilson for doing our carbon audit and multiple soil tests and bringing it to life with her enthusiasm. Thank you to the Woodland Trust for supporting us with trees. Thank you to everyone who has donated to do projects across the valley—particularly the Bray family. Thank you to Natural England and Environment Agency staff, who have made good stuff happen behind the scenes. Thank you to Paul Arkle for helping us navigate the bureaucracy required to get our farm into the best environment schemes, and for being so enthusiastic about what we are trying to do. Thank you to James Robinson for his insights into dairy farming, and the challenges of doing anything different in the current system.

Huge thanks are due to my friend Danny Teasdale, who has helped us both understand rivers better, and found the money and the diggers to get stuff done. Danny is a pragmatic conservationist of the kind we need everywhere—if you'd like to help him do conservation works in this valley, please donate to his cause (www.ucmcic.com). I am proud of our Nature-Friendly Farming Partnership and hope it goes from strength to strength.

Thank you to everyone who has volunteered to plant

trees and hedgerows. Thank you to all of our school part-
ners, who remind us all the time how lucky we are to live
here, and how much joy there is in sharing it.

Readers should also know that we are not unique or
special: this valley and the next are full of good farming
people all finding new ways to manage their land better for
food production and for nature. There are a lot of very good
farmers who care, and they give me hope.

~

The following friends kindly read the manuscript of this
book and provided helpful feedback: Nicola Wilding,
Adam Bedford, Rob Dixon, Caroline Grindrod, Kathryn
Aalto, and Patrick Holden. Thank you also to Jane Clarke
for reading the manuscript and providing incisive comments.

Any mistakes, of fact or judgment, that remain are mine
alone.

Thank you to Malcolm Maclean for letting me stay in
the "bungalow" on Uig when I needed headspace.

Thank you to Maggie Learmonth and Ozman Zafar
for being good friends and being there when needed.
Thank you to my Polish brother Lukasz for coming one
bad spring and just helping me without asking for anything in return; you picked me back up. And thank you
to Nick Offerman for showing my work respect, making
me laugh, and being a class act.

Thank you to Ian and Liz for all you do for us when things get a bit manic.

Thanks to my mum for being there and supporting us.

Thank you to my children, Molly (totally uninterested), Bea (mildly uninterested), Isaac (proud cheerleader), and Tom (completely oblivious and single-handedly responsible for about a year of delays), and my beautiful, tough, and very smart wife, Helen. Thank you from the bottom of my heart, Helen. I am so lucky to have you backing me, doing all the mundane stuff that no one celebrates but without which our life would be a shambles. None of it gets done without you shoving me forward and propping me up when the going gets tough. I love you.

Lastly, this book was inspired by two of my heroines—Rachel Carson and Jane Jacobs—writers who dared to question the "accepted wisdom" of their age and fight back against the dogmas making things worse for ordinary people. And my sincerest thanks and respect to my friend Wendell Berry, who lit a way in the dark, long ago, for us all to follow.